BRAMPTON

Produced in cooperation with the
Brampton Board of Trade

Windsor Publications (Canada) Ltd.
Burlington, Ontario

BRAMPTON
AN ILLUSTRATED HISTORY

By Helga V. Loverseed

"Partners In Progress" by
R.G. Condie and Richard Pearce

Foreword by Kenneth G. Whillans
With Contributions by Ken Moore

Dedicated to the memory of the people who made our community what it is today.

Windsor Publications, Inc.—History Book Division
Vice President, Publishing: Hal Silverman
Editorial Director: Teri Davis Greenberg
Design Director: Alexander D'Anca

Staff For *Brampton: An Illustrated History*
Editor: Lane A. Powell
Proofreader: Susan J. Muhler
Director, Corporate Biographies: Karen Story
Assistant Director, Corporate Biographies: Phyllis Gray
Editor, Corporate Biographies: Judith L. Hunter
Editorial Assistants: Brenda Berryhill, Kathy M. Brown, Susan Kanga, Pat Pittman
Layout Artist, Corporate Biographies: Mari Catherine Preimesberger
Sales Representatives, Corporate Biographies: John Bryant, Jack Allen
Designer: Ellen Ifrah

Library of Congress Cataloging-In-Publication Data:
Loverseed, Helga V., 1947-
   Brampton : an illustrated history.

   "Produced in cooperation with the Brampton Board of Trade."
   Bibliography : p.301
   Incudes index.
   1. Brampton (Ont.)—History. 2. Brampton (Ont.)—Description.
3. Brampton (Ont.)—Industries.
I. Condie, R. G. (Robert Gordon), 1931-
Partners in progress. 1987. II. Brampton Board
of Trade. III. Title.
F1059.5.B723L68   1987   971.3'535   87-6254
ISBN 0-89781-207-7

©1987 by Windsor Publications, Inc.
All Rights Reserved
Published 1987
Printed in Canada
First Edition

CONTENTS

FOREWORD 6

ACKNOWLEDGMENTS 7

INTRODUCTION 8

CHAPTER ONE • A WILDERNESS DISCOVERED 11

CHAPTER TWO • THE EARLY YEARS 23

CHAPTER THREE • INDUSTRY COMES TO TOWN 43

CHAPTER FOUR • SPIRIT, MIND, AND VOICE 71

CHAPTER FIVE • THE SHAPE OF GROWTH 103

CHAPTER SIX • CITY OF DIVERSITY 161

CHAPTER SEVEN • PARTNERS IN PROGRESS 189

BIBLIOGRAPHY 300

INDEX 301

Title page: The corner of Main and Queen streets, circa 1920. The City of Brampton grew up around this crossroads; it was originally known as Buffy's Corners after William Buffy opened a tavern on the northeast corner in 1834.

Right: Orders of dress of the Lorne Scots. Both men are in full dress uniforms, a drum major at left and an officer at right.

FOREWORD

On this special occasion of the One Hundredth Anniversary of the Brampton Board of Trade, I am pleased to extend "Congratulations" on behalf of the City of Brampton.

This anniversary publication tells the story of the history of Brampton, a history of which I am very proud to be a part. The old Ontario architecture so well illustrated in this book is reminiscent of a long-standing Ontario community with a proud past. It also presents the Brampton of 1987, a vibrant new city of opportunity, visible in the commercial, industrial, and residential architecture of today.

Brampton has always been a trading centre. One hundred years ago we had an ashery, several mills, and two busy railroad stations. It was said that on any given day long lines of horses and wagons could be seen coming to and leaving old Brampton.

The business appearance of the city has changed greatly. Years ago, Brampton was known as the "Flower Town of Canada," with rose-growing as the area's principal industry. Today we are still blooming, but in a different way. Almost 200,000 people, more than 700 industries, and several thousand businesses make Brampton the fifteenth-largest city in Canada.

People trading with people is essential to ensure prosperity in a community. Over its one hundred years of existence, the Brampton Board of Trade has played a most vital and important role in the growth and development of the city. It has served as the local voice for business, dealing with topical items of interest for the betterment of Brampton's business community and the enhancement of the quality of life for its residents. It has grown and adapted as Brampton has and continues to extend good leadership.

We offer a special word of appreciation to all those who contributed time and financial commitments to this book. I congratulate the Board on an excellent project to commemorate their One Hundredth Anniversary. I wish the Board of Trade every success for the future as their second century of service begins.

Kenneth G. Whillans
Mayor, City of Brampton

Facing page: This painting of a Peel farmhouse was completed in 1934 by Phyllis Nelson Harvey, and is now a part of the Perkins Bull Collection. Courtesy, United Church Archives, Victoria University, Toronto

Queen Street East in Brampton circa 1920.

ACKNOWLEDGMENTS

I would not have been able to garner the information necessary to write this book without the advice and expertise of a great number of people. Some took time to sit and chat about "the old days." Others referred me to various sources—poring over library shelves and rummaging through packed-to-capacity basements, to come up with research material that might help me on my way.

My particular thanks goes to Joan Vey, George Deshpande, and other members of the staff at Brampton's Chinguacousy Branch Library. A thank you also to Ann ten Cate and Diane Kuster of the Region of Peel Archives. Their help was especially appreciated because they were working in difficult circumstances. In the middle of this project they were moving from one building to another! They are now housed in the Old County Jail that created such a controversy for Brampton in the late 1800s. Despite the fact that they had barely unpacked, Ann and Diane spent hours searching patiently through their files. Diane even found time to print some photographs a couple of days before the Archives' official opening.

Peter Levick took much of the workload off me by copying and printing most of the artwork used for illustration, as did Michele Cross of Bramalea Limited, who searched her photo files and typed up a batch of paperwork. She introduced me to Stewart Davidson, Bramalea's senior vice-president, who furnished me with stacks of information about his company.

People who deserve a special mention include Bill Barber, curator of the Region of Peel Museum; Fred Kee, chairman of the Brampton Heritage Board; John Butler, president of the Peel County Historical Society; Russell Cooper, administrator of Black Creek Pioneer Village; Bob Cranch, director of Business Development for the City of Brampton; Jim Harmsworth of Harmsworth Paint & Wallpaper; and George Tavender, author of *From This Day Hence, A History of the Township of Toronto Gore.* These gentlemen took time to talk to me and fill me in on many of the "missing links."

The Brampton Board of Trade provided me with help in the persons of Nancy Lumb and her staff. Lewis Wagg, chairman of the History Book Subcommittee, was a fund of information and a terrific liaison man. A special thank you to Mayor Ken Whillans for giving me an overview of the City of Brampton today.

The guidance provided by Lane Powell, my editor at Windsor Publications, was very much appreciated. He and many others helped to bring this book alive. Finally I would like to express my gratitude to my friends. Their moral support, patience, and willingness to listen buoyed me through many busy hours.

October 1986 HELGA V. LOVERSEED

INTRODUCTION

Brampton's rise to prominence has been astonishing. At the beginning of the eighteenth century it was wilderness, known only to the Mississauga Indians, who came to the upper reaches of the Credit River to fish and hunt. By 1837, eighteen families, spurred by the promise of life in the New World, had set down roots. Today, a scant century-and-a-half later, Brampton's population has rocketed to over 180,000. City officials predict that it will easily top 300,000 before this century closes.

Its phenomenal growth can be traced to many factors, not least of which is its location. Like Mississauga, its neighbour to the south, Brampton has benefited from its proximity to metropolitan Toronto, twenty-five miles to the east. Toronto is the financial and industrial capital of Canada and modern roads connect Brampton with it and other major Canadian cities.

Highway 410, still under construction, dissects Brampton, south to north. Highway 401, Southern Ontario's "main street," is a couple of miles away. Highway 427, which runs directly to Toronto's downtown, skirts Brampton's eastern border. As well, Pearson International Airport, a ten-minute drive away, provides air links to every corner of the world.

But these are relatively modern innovations and they do not entirely explain Brampton's heady development. To understand that, we must look into the city's history.

Brampton's beginnings were humble and its progress—at least initially—was slow. By the turn of the eighteenth century, settlers from the British Isles and the United States were pouring into Upper Canada (as Ontario was then called) to settle in the Brampton area. Brampton, however, did not yet exist.

The valley of the Etobicoke River, where Brampton now stands, was described by surveyors in 1820 as low and swampy and covered with a dense, hardwood forest. Fourteen miles from Lake Ontario, it was considered then to be in the backwoods—as indeed it was.

On January 1, 1974, the Township of Toronto Gore and part of Chinguacousy Township were absorbed into what was to become the Brampton of today. Smaller settlements such as Coleraine, Castlemore, Wildfield, and Woodhill in the east, and Springbrook, Huttonville, and Churchville in the west, and Snelgrove in the north, have also become part of the City of Brampton.

In 1832 a historian noted that "there was not as much as a house in Brampton." Two years later, John Elliott (who along with John Scott and William Buffy is generally credited with having been the first settler) had laid out a village plot. By 1853, Elliott, who originally hailed from Brampton, England, and who wished to have the settlement named after his home, and a fellow settler, William Lawson (also from Brampton, England), moved to have the village incorporated.

Brampton's incorporation was the beginning of its large-scale development. An influx of immigrants in 1845 boosted its population. Within two decades, Brampton had a newspaper, the *Brampton Daily Times* (which still exists), a thriving Agricultural Society and a Mechanics' Institute. Haggert's foundry, which manufactured farm machinery, and the Dale Estates Nurseries were thriving concerns.

The laying of the Grand Trunk Railway (later Canadian National), in 1858, added further impetus to Brampton's growth. Politically and economically it was soon so important that it became the principal centre for Peel County (today the Regional Municipality of Peel). On March 29, 1873, Brampton's status as an incorporated village was elevated to that of an incorporated town. Its population, according to a Dominion census two years previous, was 2,090—a figure that didn't double until well into the 1920s. World War I and World War II put a damper on further development, but things changed drastically with the establishment of Bramalea.

Born in 1960, Bramalea was a brand-new concept, not just for Brampton, but also for Canada. A man-made, 8,000-acre development designed along the lines of similar suburbs in Europe, Bramalea, formerly part of Chinguacousy Township, was created by the giant development corporation, Bramalea Ltd. Its development was planned right from its inception, and subdivisions, parks, recreation centres, churches, schools, and industrial and commercial buildings sprang up on what was previously farmland.

Attracted by modern facilities and pleasant living conditions, companies—ranging from small service industries to major multi-national concerns—moved their factories and offices to the area. Bramalea, along with the townships, also became part of the City of Brampton in

1974.

Not everybody has been happy with the changes that Brampton has seen in recent years. When Bramalea became part of Brampton, for example, some local citizens felt they would lose their identity. But the City of Brampton is emerging with a new identity. It continues to grow at break-neck speed—its population has jumped by nearly 44 percent in the past five years, which is almost twice as fast as any other community in Canada. Future generations will know an even different city from the Brampton of today. It is impossible to stop the inevitable march of history.

The Grand Trunk Railway began serving Brampton in 1856; the station is seen here circa 1920.

The Mississauga Indians were native to the northern shores of Lake Ontario. They camped mainly along the lower Credit River, but came upriver to fish and hunt. Painting by C.A. Reid. Courtesy, Credit Valley Conservation Authority

CHAPTER ONE
A WILDERNESS DISCOVERED

The City of Brampton, twenty-five miles to the west of Toronto, is situated in the middle of the Peel plain, 715 feet above sea level, below and in sight of the Niagara Escarpment. It bustles with industry and bristles with new developments—factories, offices, schools, and subdivisions, which stretch as far as the eye can see.

At the beginning of the 1700s, it was only wilderness. The land, carved by the upheavals of the Ice Age, many milleniums before, was covered with primeval vegetation. On the highland, hardwoods such as maple predominated. Elm and other softwoods covered the lower areas. Three rivers (later to be called the Credit, Etobicoke, and Humber) meandered through the area, flowing through forests, which then were teeming with wildlife.

Before the days of roads, rivers and lakes were the major highways. The largest river (and ultimately the most useful) was the Credit. Compared to it, the Humber and Etobicoke were trickling streams. In fact, in pioneer times, the Etobicoke turned out to be more of a hindrance than a help because it flooded its banks every spring, until it was harnessed by man in 1952. The Credit proved to be the best river for navigation, waterpower, and as a food source. In autumn, its rapids and waterfalls were choked with spawning salmon.

The first people to roam the area were probably prehistoric Indians. Not much is known about these pre-Iroquoians, as they were called. They left few traces of their civilization. Occasionally farmers have found artifacts while digging in their fields, but mostly these remains belong to a later era, when Indians had come into contact with the white man.

But Brampton's history begins much later, during the time of the Mississaugas. At one time the savage Iroquois occupied the north shore of Lake Ontario. At the beginning of the 1700s, the Mississaugas, seeking more southerly hunting and fishing grounds, migrated from the southern shores of Georgian Bay and Manitoulin Island (still an Indian stronghold). They gradually drove out the Iroquois and took over the densely wooded land along the northern shores of Lake Ontario.

The Mississaugas were a band belonging to the Chippewa tribe, one of the strongest arms of the great

11

OJIBEWAY BIRCH-BARK LODGES

There were 2 Types: Angular and Domed. The frame-work of the former was made by setting poles in the ground at an angle to form a cone. The other was made by tieing branches together to form a dome. They were called Wigwams.

Both were covered with sheets of birch bark, and held down by outside poles.

Building frame of Dome shaped Lodge

The Mississauga Indians were a band belonging to the Chippewa or Ojibwa tribe. They travelled in small bands and lived in wigwams like these. Their favorite campgrounds were at the mouth of the Credit River. Courtesy, McGraw-Hill Ryerson Ltd.

Algonquian nation. The Chippewas, who are sometimes called Ojibwa (both names mean "people whose mocassins have puckered seams" in the Indian language), controlled the area north of Lake Superior as well as the region to the north and east of Lake Huron and Georgian Bay.

Like all Algonquians, the Mississaugas were a nomadic people. Their favorite camping grounds were the flats along the lower Credit River, some fourteen miles to the south of present-day Brampton, and they came up the river to fish and to hunt. Brampton's forests abounded with deer and wild animals useful for food and clothing, and ducks, geese, and wild turkeys were abundant.

The Mississaugas travelled in small bands, or sometimes single families. They lived in wigwams and ate fish, game, birds, and berries. They knew how to make maple syrup. Their campsites were littered with fish cleanings, bones, and other garbage. Their custom was to stay in a certain area until its food sources were depleted; then they would move to a new place where the woods were clean.

The river flats at the mouth of the Credit were favorite camping grounds, particularly when the salmon were spawning. The salmon came into the Credit River all the year round but they were especially numerous in the fall. This major source of food caused a constant coming and going of Indian families to the Credit. The flats became known as Indian Village (and indeed a village

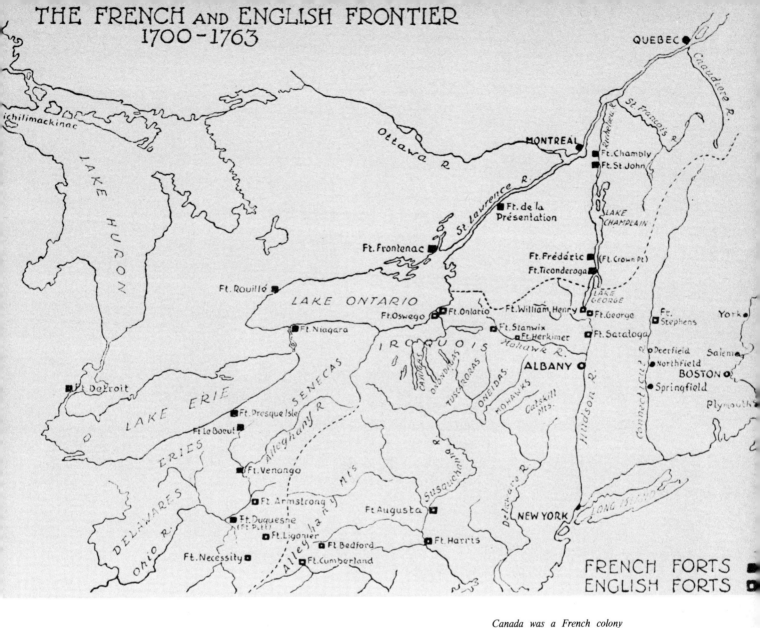

Canada was a French colony until the Treaty of Paris in 1763 passed control to the English. Ontario was at that time a part of the province of Quebec. Courtesy, McGraw-Hill Ryerson Ltd.

did spring up there in later years).

The French, who were busily employed in the Canadian fur trade, were the first to chart the area. Ducreux, the French map maker, traced the Credit River in 1660. Eight years later, Raffeux mapped it in even greater detail, and in 1757, Boucher de la Brocque, another cartographer, gave the river its name: "Riviere du Credit."

At that time, Ontario was a part of the province of Quebec in French Canada, which it remained until 1791. The French had been trading in furs along the St. Lawrence River out of Quebec and Montreal for well over a century. Their chief rivals in this lucrative trade were the English merchants operating along the Hudson River out of Albany, Williamsburg, New York, and other English colonial towns. Competition was fierce. Businessmen who controlled the trans-Appalachian trade in buckskins and furs were learning about the Canadian interior, and spurred by the thought of immense profits, they started thinking about future land speculation and development—by no means a modern concept.

Meantime, the French were pushing southwest to Ohio and the Mississippi River, in an effort to extend their fur empire and strengthen their sphere of influence. To protect their domain, they built a chain of fortified posts along Lake Ontario stretching from Fort Frontenac (now Kingston) to Fort Toronto. Fort Toronto was constructed near the mouth of the Humber River, one of

14

A WILDERNESS DISCOVERED

Facing page, top: The Mississaugas were a nomadic people, travelling in small bands to hunt and fish. A settlement evolved at the mouth of the Credit River in the mid-eighteenth century, where the Indians gathered salmon and traded furs to the French. Painting by C.A. Reid. Courtesy, Credit Valley Conservation Authority

Facing page, bottom: Brampton's early strength was its rich soil; the early farmers grew large crops of grains and vegetables. Painting by Alfred Ernest Mickle, part of the Perkins Bull Collection. Courtesy, United Church Archives, Victoria University, Toronto

Left: Peel settlers would band together for barn raisings, providing labor for a task that would be very difficult for any single family to complete alone. Painting by Owen Staples, part of the Perkins Bull Collection. Courtesy, United Church Archives, Victoria University, Toronto

A Peel farmyard. Painting by Thomas Mower Martin, part of the Perkins Bull Collection. Courtesy, United Church Archives, Victoria University, Toronto

The French traders from Quebec arrived at the Credit River in the spring. The Indians hauled in piles of pelts to exchange for guns, powder, cottons, and other trade goods. This trading continued until the late 1700s, when increased white settlement took away much of the Indians' homelands. Courtesy, McGraw-Hill Ryerson Ltd.

the major trading points. In 1749, the French military engineer Chaussegras de Lery suggested to the governor of Quebec that a trading post also be built at the mouth of the Credit River, but his advice was never followed.

Nonetheless, trading continued at the mouth of the Credit where the Indians had their encampments. The Indians, when bringing furs from the interior, preferred the deeper channels and better-beaten portages of the Credit to the Humber River or to the Etobicoke, as unreliable then as it would prove to be in later years.

The French traders from Quebec arrived at the campsites in the spring, their freight canoes loaded with goods for trading—colored cottons, guns, powder, shot, copper kettles, sashes, mirrors, trinkets, hats, shoes, blankets

—anything they thought the Indians would accept in exchange for the valuable furs. The Indians, for their part, hauled in piles of pelts—otter, mink, beaver, marten, wolf, bear, muskrat, and fox. The Indians also made many fine artifacts—intricately worked snowshoes, decorated skin jackets and pouches, belts, woven grass baskets, and moccasins adorned with bright beads and coloured porcupine quills.

Sometimes the Indians were not able to accumulate enough furs or artifacts in time for the arrival of the French, but such was the trust with which the traders held their Mississauga trappers that the French gave them the things they wanted "on credit" until they returned the following spring. It was from this practice, it is said, that the name "Credit River" was derived. It is a translation of the Indian Mes-sin-ni-ke, which in English means "trusting river."

Here, as elsewhere in Canada, the fate of the Indians was to be sealed by the coming of the white man although it took some time before the full impact of settlement was felt. In 1818 when the Township of Chinguacousy (now part of the City of Brampton) was founded, settlers were still noting that Indians roamed the land. The township's name was said to have been chosen in honor of a young Indian chief who distinguished himself as the leader of the forces that captured Fort Michilimackinac from the Americans during the War of 1812. The halfbreed son of a Scottish officer, his name is generally interpreted to mean "Tall Pines," which would indicate that his family was from the Brampton area—the alluvial soil found along the banks of Brampton's streams was a fertile bed for the growing of pines. Other sources claim that Chinguacousy is derived from "Shinwauk," the Mississauga word for "pinery."

The capture of Quebec City from the French General Montcalm by the British under General Wolfe was to drastically alter the course of Canada's history. Under the Treaty of Paris, which ended the Seven Years' War in 1763, the English fell heir to the Franco-American empire (with the exceptions only of Louisiana west of the Mississippi and the tiny islands of St. Pierre and Miquelon, off the coast of Newfoundland, which still belong to France). The British continued to trade with the Mississaugas at the Credit River but events soon overtook them. Clearly their days were numbered.

The struggle for independence by the fledgling United States of America seriously disrupted the fur trade. The English government declared that no ships other than those of the Royal Navy would be allowed to ply Lake Ontario. In 1776, when the American colonies finally broke away from England, there was an exodus of settlers, still loyal to the British Crown, who headed north, hoping to find land. For the next couple of decades, 10,000 of these United Empire Loyalists swarmed into the country, anxious to start a new life in Canada.

The British government, when they took over power in Canada from the French, had recognized the claims that the Mississauga Indians had to ownership of the land, but now they were faced with the serious problem of what to do with the influx of settlers. As well, numerous discharged soldiers sought resettlement in Canada.

In 1791 the Provinces of Upper Canada (now Ontario) and Lower Canada

In the 1790s, Ojibwa Indians were friends of Lieutenant-Colonel John Graves Simcoe and his wife, Elizabeth. Mrs. Simcoe was a prolific artist and writer, and it is from her diary that we have learned much about this early period in Canada's history. Courtesy, McGraw-Hill Ryerson Ltd.

(Quebec) were established by law. The first lieutenant-governor of Upper Canada, Lieutenant-Colonel John Graves Simcoe, arrived from England with his wife, Elizabeth Posthuma Gwillin (so named because she was born after the death of her father), and two of their six children in the autumn of the same year to oversee the newly created territory. During the next few years, the governor's family made several trips between Niagara and Toronto. Their day-to-day life was recorded in the diaries of Mrs. Simcoe, who was a prolific artist and writer.

Mrs. Simcoe admired all Indians, whom she felt had a "superior air" about them, and she enjoyed watching them perform their daily tasks. She wrote:

. . . the Gov. wishing to go some way up it (the Credit), which our Boat was too large to do, he made signs to some Indians to take us into their canoe which they did, they were two men in her which with ourselves & Sophia completely filled the canoe. They carried us about 3 miles when we came to the Rapids and went on shore. The banks were high one side covered with pine and a pretty piece of open rocky country on the other.

One of the first things Simcoe did was to declare Toronto the capital of Upper Canada and change its Indian name to the Town of York, in honor of the Duke of York (in 1834, York became Toronto again). Next he set about the

A WILDERNESS DISCOVERED

Lieutenant-Colonel John Graves Simcoe, the first governor of Upper Canada (Ontario). Courtesy, Ontario Archives

task of finding land for the United Empire Loyalists. The British had, from time to time, purchased land by treaty from the Mississaugas. One of the few portions left along Lake Ontario was the stretch between Etobicoke Creek and Burlington Bay.

This unpurchased section was known as the "Mississauga Tract," and although Simcoe had hoped that this last Indian land would not be settled for some time, he said in his writings that it ought to be set aside for the "King's Masting." This tract of land had some of the finest trees in Upper Canada and Simcoe wanted the towering pines and oaks to be reserved for the ships of the British Navy, on which English military supremacy was strongly dependent.

Inevitably, the Mississauga Tract did not remain unsettled for long. In 1793, Simcoe decided to build a military road to link the Town of York with the western part of the province. The land was surveyed in 1794 by Augustus Jones (father of Peter Jones, the famous Methodist missionary who became a chief of the Mississaugas and documented much of their life). Within two years, the soldiers of the Queen's Rangers had hacked a primitive track through the wilderness.

This east-west road, due south of the future City of Brampton, was named Dundas Street (Highway 5), after the then-Home Secretary, Henry Dundas. Little more than a narrow, stump-laden dirt path cut through the forest, it followed part of an old Indian trail that ran from Cataraqui (Kingston), around Lake Ontario to Niagara. Dundas Street was to provide passage for settlers, soldiers, and travellers through the Indian lands from Etobicoke to the "Head of the Lake" (Burlington) and would be a link with settlements west to the River Thames.

After the road was proclaimed "open for sleighs and horseback," settlers poured into the area. As yet, Brampton itself was still in the backwoods. It had no military importance and it was too far from Lake Ontario, fourteen miles to the south, to be of much significance for transportation. But the building of Dundas Street later led to the building of other major roads, which, in turn, opened up the wilderness. One such road was Hurontario (Highway 10), which runs south to north through the present-day City of Brampton. So, it was only a matter of time before settlers would push their way to the hinterland.

Two major treaties with the Mississauga Indians, in 1805 and 1818, spurred development of the area. The founding of York and the opening of Dundas Street had made the Mississauga Tract, in the words of Receiver-General Peter Russell, "indispensably necessary to connect the population of the Colony, with the seat of the King's Government."

The First Mississauga Treaty, signed on August 2, 1805, was the result of ten years' hard bargaining with Joseph Brant, the powerful chief of the Mohawks. Although born in what is now Ohio, Brant fought on the British side during the American Revolution. After the war he led his Mohawk tribe into Canada and for his loyalty was granted 700,000 acres along the Grand River near the present City of Brantford. (It is a sad reflection on the course of history that today their descendants own only 30,000 acres. They are still fighting for adequate compensation for their land and their case is dragging through the Canadian courts.)

Brant, whose Indian name was Thayendanegea ("two sticks of wood bound together"), was one of four chiefs representing the Mississaugas during the signing of the Treaty. A flamboyant character who sometimes dressed as a white man, sometimes as an Indian, he was respected by Indians and English alike. In 1776, during a visit to England, he created a sensation when he refused to kiss the hand of George III. He reasoned that he was just as much a ruler as the powerful English monarch.

After entertaining Brant at dinner, Elizabeth Simcoe wrote a detailed account of this colorful and unusual man:

He has a countenance expressive of art or cunning. He wore an English coat, with a handsome crimson silk blanket, lined with black and trimmed with gold fringe and wore a fur cap; round his neck he had a string of plaited sweet hay. It is a kind of grass which never loses its pleasant smell. The Indians are very fond of it. Its smell is like the Tonquin or Asiatic Bean.

Brant was a tough adversary, and after lengthy negotiations the terms of the treaty were finally decided upon. The First Mississauga Treaty was signed by the chiefs and Colonel William Claus, deputy superintendent general of Indian Affairs, who was acting on behalf of the Crown. A paltry £100, which even then was not a great sum of money, was paid for 70,784 acres. Moreover, most of the payment was made in trade goods rather than currency, presented once a year at the Indian Council Ring.

The land that was sold was the southerly part of the Mississauga Tract, which stretched from the Etobicoke Creek on the east, twenty-six miles westward to the outlet of Burlington Bay, and north for from five to six miles. Some land was exempted from the deal: the northern part of the Mississauga Tract (site of the future Brampton), the "flat lands" bordering the Etobicoke, a large area running one mile on each side of the Credit River, and a rectangle of property two miles long and five miles deep, which belonged to Brant himself.

The second treaty, signed in 1818, came about as an indirect result of the War of 1812 with the United States. Once again land was needed to accommodate new settlers, and it was this treaty that led, eventually, to the founding of Brampton and its neighboring townships.

Old animosities between England and the United States had remained after the American Revolution, but things came to a head when Britain, desperately fighting Napoleon, blockaded Europe and hampered the lucrative U.S. trade with France. With control of the seas her only strength, Britain imposed a "right of search" for goods that might aid the French dictator. Much to the chagrin of the Americans, the British made a habit of seizing any deserters and other British citizens they found on American ships, and pressing them into service with the Royal Navy. Exasperated, the United States Congress finally declared war on Great Britain and Canada on June 18, 1812.

For the next couple of years, Upper Canada became a major battleground. But the war finally ended and a peace treaty was signed on December 24, 1814. Later, members of the militia were given grants of land as a reward for services rendered to the crown. Once again, settlers began to surge into Upper Canada.

The government now started eyeing the northern part of the Mississauga Tract. Articles of Provisional Agreement were entered into on Wednesday, October 28, 1818. The same William Claus who had conducted proceedings for the First Mississauga Treaty thirteen years earlier once again acted for His Majesty King George III in his capacity as deputy superintendent of Indian Affairs. Representing the Indians were chiefs Weggishgomin, Cabibonike, Pagitaniquatoibe, and Kawahkitaquibe.

The balance of the Mississauga Tract, comprising 648,000 acres, was sold for the sum of 522 pounds and ten shillings. The huge tract, which adjoined the strip sold under the 1805 agreement, contained at the eastern end the balance of the future County of Peel (of which Brampton became the county town). It included the northern half of the Township of Toronto (now Mississauga), the Township of Erin, Albion, Caledon, and the Townships of Chinguacousy and the Toronto Gore, which today are part of the City of Brampton.

The next step was to measure up the land. After the Indians surrendered their hunting grounds, the government of Upper Canada set about hiring surveyors. The "New Survey," as it was called, was to prepare the land for settlers. Toronto Gore was charted in 1818. The following year, Richard Bristol completed his assessment of Chinguacousy. By October 1, 1819, the land was ready. All that remained was for the settlers to move in.

Joseph Brant, chief of the Mohawks. Although born in what is now Ohio, Brant fought on the British side during the American Revolution. Courtesy, Ontario Archives

Travel on horseback or with wagons pulled by oxen, through bush and barely-built roads, was a slow and difficult procedure. In this way, many British Loyalists made their way to Upper Canada. Courtesy, McGraw-Hill Ryerson Ltd.

CHAPTER TWO
THE EARLY YEARS

The City of Brampton today is much larger than it was in the nineteenth and early-twentieth centuries. Brampton encompasses the townships of Chinguacousy and Toronto Gore which sprang up at the beginning of the 1800s. But in order to trace Brampton's beginnings, we must look at the history of the nearby settlements as well.

Because it was in the middle of what became the Township of Chinguacousy, Brampton was among the last areas to be settled after the New Survey of 1819. Chinguacousy sprawled over 130 square miles and there was a lot of wilderness to clear. Later, during the age of the railroads, Brampton's central location was to prove a bonus.

John Lynch (who became Brampton's first reeve) gives an idyllic picture of how Brampton looked in 1820, after settlement:

Brampton was a very pretty and interesting place in the spring. The Etobicoke meandered through the streets, its banks green with leeks, but spotted with early spring flowers. There were also lots of beautiful deer in the neighbourhood, as well as bears, wolves, foxes and racoons. . . .

When Upper Canada was created, Simcoe divided it into nine districts or counties. These were further divided into townships. Peel County was first a part of the district of Nassau (later called Home District). After the Municipal Corporations Act was passed in 1849, Peel was linked with York and Ontario. Within its boundaries fell the southern part of the Township of Chinguacousy, the Gore of Toronto (also called the Toronto Gore) and the northern edge of Toronto Township (Mississauga)—all of which are part of Brampton today.

The county was split into townships because that was the unit of land division that the British administrators preferred. The government found "the settling of Planters in Townships hath very much redounded to their advantage, not only with respect to the assistance they had been able to afford each other in their civil concerns, but likewise with regard to the security they have thereby acquired against the insults and incursions of neighboring Indians or other enemies."

These townships had a certain degree of self-government, but Simcoe, who was now governor, did not

Right: After settlers had determined the layout of the land, their next chore was to set up camp and start clearing the forest. Chopping down the immense trees was a back-breaking task. Courtesy, McGraw-Hill Ryerson Ltd.

Facing page: The log house was a grander and more elaborate version of the shanty. White pine logs, mortised at the ends, were used as walls. The gaps between the logs were chinked with chips, moss, and mud. Courtesy, Region of Peel Archives

CLEARING LAND, ABOUT

like the idea of them becoming too independent. He restricted the powers of township administrators, and the only people given any authority were those who were absolutely necessary to the running of a pioneer community —pathmasters (who looked after the roads), town clerks, tax collectors, and the like. Others had minor tasks such as looking after stray animals and organizing statute labour. In pioneer days, many people paid for their land or settled their bills by working for a specified number of days instead of paying cash.

Naturally, such a paternalistic system was doomed to come apart sooner or later. As the townships developed, they demanded more control over their own affairs. In 1867 the County of Peel broke away from the counties of York and Ontario, after Brampton itself had first incorporated as a village, then as a town. A courthouse was built there, and Brampton became the county seat,

THE EARLY YEARS

a position it maintained until 1974 when the Region of Peel was created.

A year after the New Survey of 1819, settlers were making their way into the wilderness. Before being able to lay claim to the land, however, certain duties had to be performed. A fee of £5 14s. 1d. (five pounds, fourteen shillings, and one pence) was charged for every 100 acres. According to a government decree, in order to obtain a deed, at least five acres had to be cleared, fenced, and planted. It was also necessary to clear half the road allowance along the front of the property and to cut down brush and stumps to allow wagons to pass over. In addition, a cabin at least sixteen by twenty feet had to be built "in the clear" and occupied for one year. The land could not be sold for three years.

It was common practice for absentee owners to purchase land and then have somebody else do the donkey work and occupy the property long enough to obtain a patent from the government. Speculation was rife — families who had already settled in Toronto Township saw a chance to make a quick profit. They bought up bundles of land to be resold at a higher price. Surveyors grabbed up prize parcels. They considered that good land was one of the perks of their job. The Canada Company, a land promotion group, grabbed a goodly portion. The clergy and retired soldiers accounted for the rest (unfairly, ex-officers could claim up to 5,000 acres instead of the usual 200). Those not affluent enough to purchase Crown grants soon discovered that the choice plots had gone.

Nonetheless, the intrepid newcomers were determined to make a go of it. Although conditions were harsh, they were harsher still in the lands they had left behind.

Their courage and determination was daunting. To track down suitable land and inspect the various lots, they often had to make several trips back and forth from York (Toronto). Traveling on horseback or in wagons pulled by oxen through bush and barely built roads was a slow and difficult procedure.

It was also expensive. One Thomas William Magrath, Esquire, writing to

BRAMPTON

Charles Calder, who came to Chinguacousy in 1820, was among the area's first white settlers. His father noted that when they arrived, the land was still dense forest and was known as "wolf country." Courtesy, Historical Atlas of Peel County

a friend in Ireland, calculated that by the time a settler had bought 200 acres of land, purchased provisions, paid for lodging, and hired a wagon, oxen, and a guide, he could run up a bill of £178—a substantial sum for those days.

As Dundas Street had opened up the area to the south, the Gore Road and Hurontario Street gave access to the townships of Toronto Gore and Chinguacousy. In places these roads were little more than blazed trails. Travel was painfully slow and sometimes it was easier to follow the rivers and the old Indian trails. It often took several days to travel a mere thirty miles.

A Methodist circuit rider who came through Chinguacousy described what it was like to travel in these newly settled areas:

The last of August we passed into the new settlements about thirty miles from York (Toronto). English, Irish and a few American families treated us kindly. For horses there were neither roads nor feed. A pocket compass to guide us through forests of four to ten miles without a house and a hatchet to fell trees for crossing the rivers were part of our necessary outfit. . . .

Given the terrible condition of the roads, it was a wonder that there would be any settlement at all. Traveling over causeways of "corduroy," where logs were placed side by side to cover the swamps, was a bone-shaking experience. The logs were uneven and easily dislodged. One historian describes them as "looking like an inebriated graveyard." Spring floods turned these "highways" into seas of mud and at times they were completely impassable. The heavily laden wagons hauled by oxen sank up to their axles, and the animals strained to pull out of the mire. Mosquitoes and flies swarmed around the settlers. It was an ordeal for them just to reach their land.

Once there, the first step was to establish the boundaries of the property—not always an easy task, as all the new landowners had to guide them was the surveyor's blaze on the trees. Forests of matted undergrowth and fallen trees, undrained swamps and bridgeless streams, presented even more formidable difficulties.

After the layout of the land had been determined, the next chore was to set up camp and start clearing the forest. Chopping down the immense trees was a back-breaking task—especially since the axes used were little more than short-handled ship's hatchets, then in general use.

The next priority was to provide shelter for families and livestock. Labor was in short supply, and from the start, cooperation was an important part of pioneer life. There were many tasks that even with the help of grown sons and daughters could not be accomplished within one family. Neighbors helped one another in erecting houses and barns. Later these cooperative efforts evolved into "raising bees." The building of a new house or barn became a social occasion—one of the few in an otherwise work-filled life. The hungry men had to be fed, so the women went along to prepare the food. The bee gave people a chance to visit their friends and exchange a little gossip.

The first home for many pioneers was the log shanty. A crude shelter seldom more than ten feet long, eight feet wide and six feet high, it had a roof which sloped downwards towards the back. At this point it was rarely more than four feet from the ground.

THE EARLY YEARS

Above: Robert McCollum came to Chinguacousy in 1825 with his parents. He held, among other offices, those of clerk and treasurer of the Township of Chinguacousy. *Courtesy,* Historical Atlas of Peel County

Left: Peter McVean, grandson of Alexander McVean, photographed circa 1880. Alexander McVean and his four sons were the first settlers to arrive in the Toronto Gore, circa 1820. *Courtesy, Region of Peel Archives*

The log house was a grander and more elaborate version of the shanty. The usual building was about twenty feet long, eighteen feet wide, and from nine to twelve feet high. Its construction was similar to that of a shanty. Stacked white pine logs, mortised at the ends, were used as walls. The gaps between the logs were "stubbed," or chinked, with chips, moss, and mud. In some cases, settlers filled the spaces with plaster made from clay, mixed with lime or sand, when such raw materials were available.

To conserve heat and prevent drafts, early log houses were rarely built with more than one window. Sometimes they had dirt floors, although later they were planked. They often remained in an unfinished condition for several years,

THE EARLY YEARS

because pioneers, concerned only with basic shelter, did not find time to finish them until other more essential tasks were done.

Mrs. Susanna Moodie, sister of Samuel Strickland who was an employee of the Canada Company, did not mince words when it came to telling people what she thought about such homes. In *Roughing It In The Bush,* published in 1852, she describes them as "dens of dirt and misery, which would in many instances, be shamed by an English pig-sty."

As the townships developed and the settlers became more prosperous, their homes improved. Rooms were added when children came along. Walls were whitewashed (paint was scarce). A stoop or a veranda was built along the front. But compared to Britain or the United States, the average house in Upper Canada was considered only basic. As late as 1834, assessors calculated that 23 percent of the houses were still log cabins.

The first settlers to arrive in the Toronto Gore (thus-named because it was a wedge-shaped piece of land between Chinguacousy and the townships of Vaughan, Etobicoke, and Toronto) were Scots. Just after the Gore was surveyed, Alexander McVean, with his four sons and daughter, set off along Dundas Street with their mare, Meg. It is believed that they headed north on an old Indian trail along the Humber River.

McVean purchased 200 acres on Lot 7, Concession 8, from Reuben Sherwood. Later he purchased another 200 acres in Lot 6. Soon he had cleared fifty acres of Lot 6—the largest cleared area in the Gore at that time—and after he had prepared the land, he erected a large log home. He built his home on the banks of the Humber River about a quarter of a mile above the present-day Highway 7, near the road that bears his name, and in deference to his native land, he named it Torbulglen. This tendency to name homes and hamlets after the places left behind was common among pioneers—perhaps because it alleviated their homesickness. In the Toronto Gore, which was settled largely by the Irish, there were at least three villages with Irish names —Tullamore, Castlemore, and Coleraine.

Like many settlers, the McVeans had their share of hardship. Tragedy struck in 1822, when a son, Alexander Jr., became ill and died. (Another son, Archibald, went on to become Director of the Township and of the County Agricultural Society.) In December of that year, their house was devastated by fire. After appealing to the government for aid, McVean was given a further 200 acres of land. His home was rebuilt by neighbours.

The completion of the Gore Road in 1822 prompted further migrations to the wilderness. The McVeans were quickly followed by the Grahams, Bells, Lawrences, and Philipses. Elisha Lawrence, from New Brunswick, paid for his land by helping to build the road. In return for services rendered, he was given Lot 2, Concession 9.

The major concern of a pioneer farmer was to grow enough food for his family, with some left over to barter at the store or sell for cash, which was needed to pay taxes and other expenses. Few farms were without a patch of flax. Sheep provided wool. Wheat provided flour and an orchard became part of almost every homestead, as did a garden planted with vegetables and herbs.

But as the communities progressed, it became obvious that there was a need for facilities other than those provided by the family farm. There was

Facing page, top: This is the farm belonging to George Bland. The Bland family were settlers in the Toronto Gore in the 1870s. John Bland, a Yorkshireman, served as the township's clerk. Courtesy, Historical Atlas of Peel County

Facing page, bottom: Archibald McVean (son of Alexander) and his wife, Helen Gordon, came to the Toronto Gore shortly after the survey in 1819. Archibald was a director of the county Agricultural Society. Courtesy, A History of Peel County

THE EARLY YEARS

no grain mill. The *Historical Atlas of Peel County,* published in 1877, noted that when Elisha Lawrence first settled in the Gore, "he had to carry his grist fifteen miles to Richmond Hill." But the water table within the Gore was high, with many swamps and bogs, and it was only a matter of time before the waterpower was harnessed. The enterprising McVeans agreed to build a grist mill. They purchased Lot 5, Concession 8 from the Anglican Church—a point where the flow was swiftest. By November 1834, the mill had begun to operate. It was an instant success and the McVeans continued to operate it for the next eleven years.

At the eastern edge of the Gore, hamlets such as Castlemore, Coleraine, and Ebenezer were slowly taking shape. Ebenezer was little more than a church and a graveyard, but in pioneer times even such small settlements were thriving concerns. When the railroads came to Brampton some hamlets died away. Today, many are mere names on the map.

By the middle of the nineteenth century, Castlemore had a hotel, a church, a post office, a blacksmith's shop, and a school. Although still in its infancy, a school system was evolving throughout the newly settled areas. The Common

Above: In the 1870s, this large farm and house—Rose Villa—belonged to James Cooke. The Gore's first farmers made good use of the area's fertile soil, growing grains, vegetables, and fruit. Courtesy, Historical Atlas of Peel County

Facing page: An early water-powered gristmill of the type found in Upper Canada in the early 1800s. Oxen were used to pull the heavy wagons or sleds loaded with grain. Courtesy, McGraw-Hill Ryerson Ltd.

THE ARCHDEKIN FAMILY

The Archdekin family may not have been raised in Brampton, but the impact they have had on the community has been as great as any native Brampton clan. While one son dedicated his life to the betterment of the town in the political arena, the others in the family lent their business skills to Brampton.

Originally, the Archdekins settled in the Mayfield area, which is today part of the City of Brampton. Their farm was located on the 4th Line Chinguacousy and the 17th Sideroad. Peter and Prudence Archdekin had come from Ireland about 1820, and a brother of Peter's had come with them, but had opted to settle in the United States. Peter and his wife had nine children; large families would become a trend in later generations.

The second-eldest son, Thomas, decided to leave the farm and open an establishment called the Bay Horse Inn. However, Thomas sold his hotel business to Mugucks, opened a general store on the southeast corner, and eventually moved to the farm where he built a house, which still stands today on the corner of Dixie Road and 17th Sideroad. This was about 1875.

The family remained in the Mayfield area until Thomas's great-great-grandson, Stan, left the province and moved to Saskatchewan. He arrived in Moose Jaw on what was called a harvest excursion, where people came from all over to help with the harvest. While he was there, Stan met and married Muriel Harris and the couple had four children: Elmore, Evelyn, Viola, and Jim.

By 1927 Stan felt the urge to be back with his kin and he returned to Mayfield and busied himself with mixed farming. He and Muriel had five more children: Fennell, Albert, Leo, Marlene, and Charlene.

Elmore, the eldest of Stan's children, was the first to make his way to Brampton. In 1941 World War II was two years old and Elmore joined the army. However, a training accident forced him to be discharged and he looked to enter a trade.

In 1943 Elmore entered the plumbing business in Brampton, which remained his livelihood until 1978. The business is now run by his son Elgin, who trained under his father. Elmore remembers that it was by accident that plumbing became his life's work. He had originally intended to enter munitions work, but a friend suggested plumbing as the way to go, and it was the right choice.

Elmore was also one of the charter members of the Brampton Flying Club in 1946. At first, the club's Tiger Moth was kept at Stan's farm until a suitable landing site could be found. According to Elmore, the club was more of a social club than anything else, with members going flying instead of playing golf.

For twenty-five years, Elmore was a member of the Brampton Hydro Commission. He served as chairman for seventeen of those years. For seven years he was a parole officer with the Ontario Ministry of Correctional Services.

Jim Archdekin followed his brother into Brampton and for twenty

years operated a cabinet-making business. He is probably better remembered for his long service to Brampton first as a member of town council and then as mayor. Jim often said that his greatest thrill in office was the opportunity to welcome the Queen during her visit to Brampton for the city's centennial celebrations.

He served as a popular mayor of Brampton from 1969 until his death in 1982. More than 3,000 people paid their respects in perhaps the largest funeral the city has ever seen.

Albert Archdekin went into business with Jim in 1951 and the two worked side by side until Albert went on his own. Albert is now assisted by his sons, Peter and Ken.

According to Albert, none of the family wanted to follow their father into the farming business, and since Brampton was like home to them, they eventually all moved here. Fennell Archdekin was a doctor and coroner in Muskoka.

The youngest son, Leo, entered the funeral business soon after Stan's death in 1956. He eventually opened his funeral parlor on Main Street North. He has since sold the business, but the name remains as a reminder.

Even though the Archdekin family led diverse lifestyles when they came to Brampton, they all had one thing in common—they had success as a common denominator.

—Ken Moore

Front (l-r): Fennel, Leo, Jim, and Albert Archdekin. Rear: Elmore and Stan Archdekin

THE EARLY YEARS

School Act of 1816 "gave authority to the inhabitants of any township to unite and build a school, provided they furnished at least twenty pupils and paid part of the teacher's salary." As the settlements became more prosperous, money was found to hire instructors for the children. Castlemore's school was built on the property of Patrick Dougherty on Lot 12. Theophilis Norton was the first teacher.

The Anglican Church had a strong following and many of the residents, including the Kerseys and the Blands, held garden parties and suppers in order to raise funds. It was a reflection of the cooperative spirit of the pioneers that these events were supported by both Catholics and Protestants.

On the western side of the Gore, along the Sixth Line (Airport Road) were hamlets like Woodhill, Stanley Mills, and Tullamore. Grahamsville, at the

Facing page: This woman is boiling potash—a valuable commodity in pioneer times that fetched from £2 to £6 a barrel. Brampton's first business was believed to be an ashery run by John Scott. Courtesy, McGraw-Hill Ryerson Ltd.

This mid-1850s Brampton street scene (Queen Street West) shows early retail stores, including William Golding's bakery, T.D. Shenich's shoe shop, and C. Stork's drug store. Courtesy, A History of Peel County

northern edge of Toronto Township (it later became part of the City of Brampton), was founded by the Grahams who arrived in the area shortly after the McVeans. The first store was operated by George and Thomas Graham, who had been granted a tavern license as early as December 1819. The Magnet Hotel was built in 1831. Later, Grahamsville became a meeting place for municipal officials. Township business was conducted at Graham's Inn.

Not suprisingly, many of the early pioneers became involved in local politics. They were not only farmers, but businessmen and administrators as well. Simon Grant was the Gore's first township commissioner, serving from 1832 to 1846. He was followed by James Sleightholme, an immigrant from England. Sleightholme served as commissioner until an 1849 by-law allowed townships to set up local councils. He became the Gore's first reeve. John Bland, a Yorkshireman, served as township clerk.

The Gore's rich, loamy soil soon won it a banner reputation. It was very fertile, and even today there are several farms in this corner of Brampton. Much of the land remains green. Between the Gore and the residential and industrial developments of Bramalea lay several parks and recreation areas—Castlemore Golf and Country Club, Woodlands Golf and Country Club, and the Claireville Conservation Area, through which flows the Humber River.

In 1849, John Lynch (later Brampton's first reeve) reported that the average wheat yield in the Gore was 21.7 bushels per acre—one of the largest in Upper Canada. Brampton land prices were the highest in Peel County. Uncleared plots sold for £5 13s. per acre; cleared land was worth £6 13s. 3d.

Meanwhile, in the rest of Chinguacousy (the Gore was part of the Township of Chinguacousy until 1831), settlement was developing apace. Hurontario Street (Highway 10), which ran from the south in a northwesterly direction, had been built to connect Port Credit with Collingwood and Georgian Bay and it crossed the middle of the township. The New Survey adopted Centre Road (as the part running through Chinguacousy was called) as the control line, and six concession roads were laid out on either side. Today, Centre Road is Brampton's Main Street.

Settlers flooded in, but it took a while to clear the land. Dense bush still covered vast areas of the township. The *Historical Atlas of Peel County* relates how Charles Bowles (father of Thomas Bowles, reeve of Chinguacousy) travelled from Toronto to Chinguacousy in 1827: "He hired a team and waggon thinking he could get through with them, but was compelled to leave them before arriving at the end of his destination and make the best of his way on foot through the trackless forest."

Wolves and other wild animals still roamed the forest. While they were a source of food for the newcomers, they also wreaked havoc with the livestock. In the records of the township meeting for 1829, Chinguacousy's first Township Clerk, John Scott, wrote: "It was passed by a vote of the meeting that sixpence per head shall be paid for wolves killed within the Town, which shall be collected by the Collector."

By-laws dealt with the need to protect domestic animals and prevent them from straying. One of the first read: "Pigs under 30 weight are not to be considered free commoners & if they are allowed to stray and injure a neighbour, the damage must be assessed by two interested persons and the owner of the pigs must pay the damage."

As the wilderness receded, the game disappeared and with it went the Mississauga Indians. They had continued to hunt in Chinguacousy for some time after the Treaty of 1818. To a people not familiar with the concept of owning land, the notion that they could no longer use it was quite foreign to them. Sale or no sale, they reckoned that it was still theirs for use.

At first, the settlers didn't mind their presence—probably because there was enough land for everybody. In fact they often hired Indians to protect their property and animals. But as more people moved in, the Indians' presence was resented. The Mississaugas, in any case, were dwindling in numbers. They had no resistance to the highly contagious diseases imported by the white man.

THE EARLY YEARS

This Chinguacousy farmhouse was built in the 1830s during the years when the area was first settled. It was located near Lot 8 on the Fifth Line (Torbram Road) and was saved from demolition in 1973. It is now called "The Doctor's Home" and is part of Black Creek Pioneer Village in North York. Courtesy, Metropolitan Toronto and Region Conservation Authority

Many fell prey to consumption, pleurisy, measles, smallpox, and whooping-cough. Infant deaths were common.

The Reverend Peter Jones (Kahkewaquonaby), a chief of the Credit River Indians and the half-breed son of surveyor Augustus Jones, was a Methodist missionary who recorded the day-to-day lives of the Missisaugas. In his writings he bemoaned the plight of his people: "How the scene has changed since the white man discovered our country!" he wrote in 1834. "Where are the aborigines who once thronged the shores of the lakes and rivers on which the white man has now reared his dwelling and amassed his wealth? The red man is gone, and a strange people occupy his place. . . ."

By 1821, the Township of Chinguacousy had a population of 412, with 230 acres of cultivated land. In the next fifty years the total developed acreage had grown to 80,271 and the population to 6,129. By the 1830s, bridges had been built over the rivers, and the roads were gradually improved to allow the

IMMIGRATION

In recent years one only has to look around Brampton to see the changing face of the community—a greater influx of minorities are integrating themselves and changing the cultural fabric of the city.

The areas around Toronto, and especially Brampton, have become popular with these newcomers to Canada. There are several nationalites that make up more than 3 percent of the population, including Dutch, Italian, Portuguese, German, Indian, and Pakistani.

In Brampton, the Italian community represents a significant percentage of the population, which places them as Brampton's largest minority group. Italian immigration to Canada began in earnest at the end of the nineteenth century, but was curtailed prior to World War II as fascism became rampant in Italy. After World War II, there was a great influx of immigrants from the Netherlands who specialized in the flower industry. Immigration increased in the 1950s when workers for the construction trade were needed. Today about 70 percent of the construction industry in Ontario is managed by Italians.

The majority of Italians that come to Brampton are from the "Little Italy" areas of Toronto, and most of them have been in Canada for at least fifteen years. Numerous cultural organizations in the Region of Peel have assisted the Italian community in merging into local life, and their involvement has enriched Brampton as a community.

Portuguese immigrants make up almost four-and-a-half percent of Brampton's population. The first attempt to organize Brampton's Portuguese community came in 1968 with the formation of the Olympic Portuguese Club. This group lasted only three years, but was reorganized as the Academiade de Brampton, and in 1977 it was renamed the Academia de Brampton. With more than 300 members, the association is concerned with the organization of cultural and recreational activities. Classes in Portuguese were started in 1974 by Evangelista de Oliveira and his wife at St. Mary's and St. Anne's schools. The growth of the Portuguese community is evident not only in the growth of cultural organizations, but also in the development of many business ventures around the city.

Brampton has a large German population. The German Canadian Club, Hansa Haus, which is located on Highway 10, boasts Canada's only German heritage museum.

Relative newcomers to Brampton are immigrants from India and Pakistan. According to the Peel Multicultural Council, most East Indians come to the area without sponsorship, relying instead on their qualifications for admission. Although the term "East Indian" suggests a homogeneous group, in reality this ethnic group is composed of people who speak one or more of fifteen different languages and 400 dialects, in addition to English. As well, there are seven major religions practiced by the Indo-Pakistani groups. One of these, the Sikh religion, calls for its male adherents to wear their hair tied up in a turban, thus making them a very visible minority.

These and many other minority groups make very special contributions to the city, making Brampton a more interesting place to live, and continuing to change it and make it more vital with each new arrival.

—Ken Moore

This was the residence of William Elliott, the son of John Elliott —believed to be one of the two men who gave Brampton its name. William Elliott operated a music store on Queen Street East. Courtesy, Historical Atlas of Peel County

marketing of grain and other farm produce.

As was the case in the Gore, hamlets developed around the junctions of roads. Springbrook, home to fifty souls, grew up along the Third Line in west Chinguacousy (Creditview Road). It boasted a couple of churches and a temperance hotel. By 1847, it even had a school, albeit only a crude log building, which looked out onto a field filled with stumps. The settlers had only just cleared the land.

South of Springbrook was Churchville on the northern edge of Toronto Township. It occupied a prime position on the Credit River, and during the first half of the 1800s, it was a thriving community, with several grist mills, saw mills, stores, and a hotel. It became the centre of much excitement on August 15, 1837, when William Lyon Mackenzie, the leader of the radical wing of the Reform Party, held a meeting there to enlist volunteers to ride with him in a rebellion. The township had a strong contingent of Orangemen, many of whom sympathized with the rebel leader, but they disapproved of Mackenzie's extremism, particularly when he talked about breaking away from the British Empire. Six hundred people turned out for the rally but there were unconfirmed reports of a conspiracy to assassinate the fiery speaker. The meeting ended in a violent brawl but Mackenzie was spirited away to safety.

Between Springbrook and Churchville lay Huttonville, originally known as Wolf's Den. Situated west of the Fourth Line, it had, like Churchville, good access to waterpower (the Huttonville dam later provided Brampton with electricity). A saw and turning mill was built in 1855 by town founder J.P. Hutton.

As the settlements became more established, the focus shifted to Brampton. A small inn operated by Martin Salisbury (one mile from what is now Brampton's Four Corners) became a centre for farmers and tradesmen who came here to chat and do business. According to an 1827 assessment, Salisbury himself owned no land (it belonged to Archibald Pickard) and the only livestock he possessed was one horse and a cow. The inn, however, became a popular gathering place, and it was the site of a regular market fair.

Brampton proper appears to have started around the "Four Corners." In 1834, William Buffy opened a "pretty respectable tavern" on the northeast corner of Queen and Main, and before too long, it became known as "Buffy's Corners." Judge John Scott started Brampton's first ashery. Potash, a valuable commodity in pioneer times, fetched from £2 to £6 a barrel. Scott also opened a small chopping mill and a distillery. His mill, it is said, attracted a great deal of attention. Curiously, the stones operated vertically instead of horizontally.

Samuel Kenney was a speculator who bought land (Lot 5, lst East) merely to sell it at a profit. Early in the 1820s, he sold it for £100 to John Elliott, a settler from Brampton in Cumberland, England. Elliott belonged to the Primitive Methodist Church, a puritanical sect whose trademark was fervid, evangelical preaching. Sometimes they were called "Ranters" because their services were characterized by fiery sermons during which the congregation joined in with loud "Amens" and "Hallelujahs."

The upright Mr. Elliott took a dim view of the inn at Buffy's Corners. Concerned that the fast-growing settlement was being named after such an "unsavory" establishment, Elliott got together with a fellow Englishman, preacher, and merchant named William Lawson, and set about changing the settlement's name to Brampton, in honor of their home town. Who actually came up with the idea first is a matter of dispute, but two almost identical plaques in the Primitive Methodist Church in Brampton, England, attributes the naming to both men; one is inscribed with Elliott's name, the other with Lawson's.

At any rate, both Elliott and Lawson and their descendants went on to play an important role in the building of the town and indeed the country. Richard Elliott, John's son, owned a repair shop on Brampton's Queen Street East. His brother, William, kept a music store on Queen Street West. William Lawson, who later moved to Hamilton, became a justice of the peace. His great grandson, The Honorable Ray Lawson, O.B.E., LL.D., D.C.L., served as Lieutenant Governor of Ontario from 1946 to 1952.

Thanks to the Elliotts and the Lawsons, Brampton became the heart of Primitive Methodism in Canada. It appeared as a stopping place on the Primitive Methodist circuit—an involved schedule made necessary in those days because of the scarcity of ordained ministers—and it became the site of the first annual conference of the denomination.

Although they were deeply religious, the Primitive Methodists knew how to turn a buck. They combined business acumen with shrewdness and a capacity for hard work. By the end of the 1830s, Brampton was a hive of activity. It was in the heart of what had become one of the richest agricultural areas in Upper Canada. A decade later, a branch of the Home District (Peel County) Agricultural Society was holding semi-annual fairs and annual ploughing matches, at which merchandise was traded and farmers' skills were tested.

By the beginning of the 1850s, Brampton's population had jumped to 550 from only 18 people in 1837. It was time for the settlement to seek a change

THE EARLY YEARS

The minutes of the first meeting of the Municipal Council of Brampton, which was formed following the settlement's incorporation as a village in 1853. Courtesy, Historical Atlas of Peel County

in its legal status. On Monday, January 17, 1853, Brampton was incorporated as a village—a move that was eventually to give it more political power and a say in the running of its own affairs. The village councillors, sworn in by W.B. Reeve, Esquire, were John Elliott, Peleg Howland, John Holmes, and John Lynch. Lynch was appointed Brampton's first reeve.

Lynch served as a justice of the peace for the next fifty years. Peleg Howland became one of Canada's most outstanding merchants, and his brother, Sir William, went on to became lieutenant-governor of Ontario in 1868. Kenneth Chisholm was the municipality's first county warden. Later he became an MPP (Member of the Provincial Parliament) and sat in the Ontario legislature throughout the 1870s and 1880s.

Such men, who combined entrepreneurial skills with political ability, were the backbone of Brampton's early days. In more ways than one, they helped to put it on the map. Brampton's leaders envisioned a prosperous future for their newly incorporated village. They were not disappointed.

Men cleaning Queen Street East circa 1900 in front of the Dominion building and the old Queen's Hotel. Note that even at this time the road was little more than a mud track. Roads were not macadamized till the advent of the automobile. Courtesy, Region of Peel Archives

CHAPTER THREE
INDUSTRY COMES TO TOWN

The decision to incorporate Brampton as a village in 1853 was the result of many factors—not least of which was the arrival of the railroad. Brampton's central location, which had proved a setback in pioneer times when land had to be cleared, was now seen as an advantage. If the railroad was brought to Brampton, it was argued, it would open up Chinguacousy and the Gore to large-scale development. Brampton's indefatigable movers and shakers made sure the railway companies would route their railroads through the centre of town, and by the end of the nineteenth century, Brampton had not just one railroad, but two.

The 1850s ushered in a boom. Farmers now had a means of distributing their goods to other parts of the country (and ultimately overseas) and the travel time from Toronto was cut from two days to a mere two hours. Improved communication and transportation gave a terrific boost to the community and Brampton became the archetypal prosperous Victorian town, replete with entrepreneurs and bustling with factories and mills.

Booms and depressions tend to go in cycles, and by the end of the nineteenth century, Brampton had hit a slump. Good land was at a premium. Prices had shot up and much of the labor force—transient at the best of times—had moved on to find work in the prairie provinces and more prosperous communities. The prairie provinces not only attracted manpower, but were producing crops in enormous quantities. Their capacity for growing wheat proved to be much higher even than Peel's, which till then had been in the forefront of the farm business.

As the markets for their crops shrank, farmers, and consequently farm implement manufacturers, went out of business. Brampton's economy suffered, but in time, through the efforts of its businessmen and municipal leaders, other industries took the place of agriculture and its supporting businesses. This enabled Brampton to weather the economic vicissitudes which followed in the 1890s and throughout the twentieth century.

By 1914, Brampton had good roads, gas, a dependable water system, electricity and telephones. The first "horseless carriage" had made its appearance. Those innovations laid the groundwork for the next fifty years. Brampton didn't boom again as it had done in the mid-1800s and early 1900s until major developments commenced in the 1950s.

Curiously, the decision to bring the railroad to Brampton came at a time when the roads were actually being improved. By 1853, the Hurontario Plank Road Company had planked Brampton's main street all the way from Port Credit to Edmonton (Snelgrove). A toll was levied to pay for its upkeep. To ensure that it was maintained in good condition, all able-bodied males were required to contribute a certain number of man hours.

Unfortunately, Hurontario Street didn't stay in a usable condition for long. Wagons quickly wore through the planking, which snapped or sank into the mud. Every time there was a storm, Brampton was cut off from the outside world. Stage coaches had a terrible time. The Orangeville and Brampton Stage Line, owned by a Mr. J. Lewis, ran a regular service whenever it could, but even it wasn't able to keep its coaches running in wet weather, which was particularly prevalent in spring and fall.

An amusing anecdote by Daniel Pratley, who was interviewed by *The Conservator* in 1942 when he was 100 years old, gives some idea of what Hurontario Street was like:

I used to drive loads of wheat over it to the ships. It was the only road like it I had even seen; miles and miles of sturdy planks running all the way from here to the lake. It was breaking up by the time I got here and I often took broken planks home for the fires.

The businessmen who ran Brampton realized that in order for the village to grow, it could not rely on its roads. Clearly a railroad was the next logical step.

Railroads were all the rage in the mid-1800s, and for a while private enterprise went crazy. Caught up in the enthusiasm of the moment, companies applied for charters as if there were no tomorrow. No fewer than fifty-six grants were given out in less than three years. In their rush to capitalize on this new form of transportation, companies seldom bothered to calculate whether or not their ideas were economically feasible. Inevitably some never got beyond the charter stage. Others made huge profits. Many went bankrupt.

Railway fever hit Brampton in 1852. Promoters of the Toronto and Guelph Railway (later the Grand Trunk Railway) made a sales pitch to John Lynch, then reeve of Chinguacousy, and persuaded him that a railway company would greatly boost the development of the village. Convinced that they were right, Lynch called an emergency meeting of the Chinguacousy Council. He asked that the usual rules of business be suspended to expedite the passage of a by-law that would allow Chinguacousy Township to take stock in the proposed Toronto and Guelph Railway. The company had promised that the railway would pass through Brampton and that there would be "a Depot or Station erected in the said village." The amount of capital needed was £10,000. Council agreed.

Given the uncertainty of such a venture, it is amazing that a conservative community such as Brampton agreed to take the plunge. It is perhaps to their

INDUSTRY COMES TO TOWN

Richard Elliott and his wife, Mary Day, with their three sons —Walter, Frank, and Harry. Richard was the son of John Elliott, who, with William Lawson, founded Brampton and named it after their hometown in England. Courtesy, Region of Peel Archives

eternal credit that they did. It was obvious a railroad would benefit the community but village officials still took an enormous risk.

£10,000 was a horrendous amount when compared to other values of the time. In 1855, for example, the total budget of Chinguacousy Township amounted to only £436 5s., over a third of which was raised from the license fees of twenty-one taverns. Obviously the township politicians felt that having a financial stake in the railway company was the one way that they could ensure that the line ran through the centre of Chinguacousy. So keen were they to accommodate the railroad that they even agreed to rearrange Brampton's streets.

Some members of Council reasoned that even if all the promised funds could not be raised, it would be too late to change the railway's route after the lines were laid. As it happened, Chinguacousy never had to underwrite the venture. Before the line was complete, the Toronto and Guelph Railway was taken over. The Canada Grand Trunk Railway Company bought it out. Grand Trunk paid off the municipal stock issue and local backers were off the hook.

Once the plans to build the railroad had been approved, everybody wanted to get in on the act. There were endless arguments over where Brampton's

When the railways came to Brampton, its future was assured. The Grand Trunk Railway, which became part of the Canadian National (C.N.R.), opened on June 16, 1856. This is the C.N.R. station as it looked in 1890. Courtesy, Brampton Public Library, Chinguacousy Branch

railway station should be built. Major George Wright, an influential grain merchant and politician, lived on the north side of Church Street West and he wanted to sell off some land. Other council members preferred a site on John Elliott's property on the south side of the tracks at the intersection of Queen Street East. The railway company, tired of the constant bickering, threatened to pull out and abandon Brampton altogether—an idea that had merit in some quarters, as at that point it still looked as if Chinguacousy would have to foot the bill.

Finally, Wright, a forceful, dynamic man, who was used to getting his own way, pushed the sale through. His 11.6 acres went for the princely sum of £932 —a huge amount for land in a fledgling village. Wright's dealings in land and grain contributed to Brampton's prosperity in later years, but he amassed a considerable fortune in the process.

On June 16, 1856, the Grand Trunk Railway opened in a blaze of glory. Bramptonians could now travel to Toronto in a couple of hours. They could also ship their agricultural produce to Toronto's harbor and from there it could be shipped to every corner of the world. The railroad brought prosperity to Brampton but inevitably it sounded the death knell of some of the smaller hamlets. The outlying villages of the Gore and Chingaucousy now became isolated and off the beaten track.

The railroad spurred Brampton into a fever of activity. Opportunists poured into town. Over the next five years, its population almost tripled. Not surprisingly, the boom attracted its share of scoundrels. Wesley Todd, a self-proclaimed "merchant," managed to inveigle $80,000 from the Bank of Montreal. After it was discovered that he was forging the signatures of innocent citizens, he was forced to skip town. The law caught up with him and he spent

INDUSTRY COMES TO TOWN

James Golding (holding reins) with his wife and mother, in front of his bakery on the corner of Main and Queen Street West (the old Four Corners area). Golding, seated here in his Gladstone horse buggy in the late 1800s, moved his bakery around the corner when the Bank of Montreal purchased the Four Corners property on which his bakery was located. Courtesy, City of Brampton

seven years in a Kingston prison.

More typical was William Mahaffy, who arrived in Brampton in 1856. A carriage maker by trade, he set up shop and by 1870 was one of Brampton's wealthiest men. Before too long Brampton had a brewery, a blacksmith, saddlers, a tannery, and several taverns. There were also lawyers and doctors—among them, Dr. Alexander Pattulo (later editor of the *Brampton Times*) and Dr. David Heggie (later surgeon of the County Jail).

As was the case in the rest of Canada West (as Upper Canada was now called), immigration, settlement, and population growth was having its effect on the once-agrarian frontiers. Good land was becoming scarce and highly priced. The nature and size of families, the opportunity for better education (students could now commute by train to the University of Toronto), and geographical mobility were changing the face of society. Brampton was moving out of the pioneer era.

Farming was becoming more sophisticated, and as it moved from subsistence level to a major industry it created a demand for technology, supplies, improved transportation, and power. In 1853, an act of Parliament was passed which empowered counties to form boards of agriculture to help educate farmers and to provide funds and expertise. A government grant was provided to maintain such societies and that year seventy-five farmers and their supporters got together to form the County of Peel Agricultural Society.

The formation of such a society gave farmers a forum for keeping up with the latest developments in breeding of livestock and the growing of crops. Peel's first county fair was held in Brampton that fall, at the corner of Main and Queen streets. Today the Brampton Fair is a major annual event that attracts dozens of exhibitors from all over the country. In 1853 it was a very modest

Established in 1849, Haggert Bros. made farm machinery like the Brampton triple harvester, steam engines, and boilers. Their machinery won awards at international exhibitions and helped to industrialize Brampton's farm business. Courtesy, Brampton Public Library

affair. Grain, vegetables, and dairy products were laid out on tables. Horses and cattle were shown on the road.

The main objective of the Agricultural Society was to enable wealthy farmers to help the less well-to-do. Smallholders who could not afford to purchase large quantities of seeds and crops were able to obtain them from the Society, which, because it had so many members, was able to buy in bulk. As well, the Society could bargain for good breeding bulls and stallions. In the early days, a great deal of emphasis was placed on the improvement of the draft horse, then the major source of power.

By banding together in this manner, the farmers were able to escape from the restrictions of self-sufficient farming. They now had a market outside the local area and they were in an ideal position to take advantage of their situation. The Crimean War (1853-1856) created a high demand for wheat from North America. It also brought record-high prices. The buoyant market coupled with more efficient planting and harvesting methods allowed farmers to work faster

INDUSTRY COMES TO TOWN

with fewer men. Horsepower had replaced manpower for most major chores.

The swing from subsistence to commercial agriculture was stimulated by the application of animal power to mowers and reapers, and as farming developed, it became more mechanized. There was a demand for services, and goods like nails, horseshoes, and other farm necessities. Blacksmiths' shops did a roaring trade. In 1849, the Haggert brothers' foundry opened for business as a farm equipment manufactory. Theirs was the first major industry in the area and for the next forty years they were Brampton's largest employer.

With Brampton able to service the industry and farmers of Chinguacousy and the Gore, suggestions were made that Peel should become a separate county and that Brampton should be its county town. A courthouse and a county jail were all that was needed, and given that Brampton had become so prosperous, it seemed the obvious choice.

For a while, things got pretty hot on the political front. The District Councils Act of 1841 had established the first real municipal self-government in Canada, and the incorporation of Brampton as a village in January 1853 gave the County of Peel in general, and the Township of Chinguacousy in particular, more political clout. The central location of Brampton, its railway (which connected to a network of others), its incorporated position, and its growing wealth made it a strong contender for county "capital." Of course, not everyone wanted to see Brampton get ahead. Villages such as Streetsville and Malton (now in the City of Mississauga), which had been hives of industry in pioneer times when Brampton was still a rural crossroads, strongly resented what was to them an upstart community.

The first step was for the County of Peel to break away from the counties of York and Ontario. In 1856, a separation by-law was passed. But that was only the start. Years of political ping-ponging followed while municipalities debated which village was to become the county town. At a January 1858 meeting, Major George Wright and Isaiah Faulkner of Caledon made the motion "that Brampton be hereby declared the County Town of Peel." The motion was declared carried, but no vote was recorded. Apparently only five members were at the meeting—not enough for a quorum on the ten-member council.

The following year the vote went to Malton. That decision was overruled. Next, the reeves and deputy reeves considered Malton, Streetsville, and Brampton. Brampton came out on top but the vote was lost when money to build a courthouse and jail could not be raised.

The shenanigans that went on at these meetings can only be imagined. Suffice to say that there was much heated debate. Over a period of two years councillors were called upon to vote on the "county town" issue no fewer than 160 times. Dozens of places in every township were proposed as suitable sites. Malton and Streetsville were joined by Churchville, Silver Creek, Cheltenham, Tullamore, and Cooksville—all felt they had a valid claim.

Finally the Township of Chingaucousy gave Robert Smith and Michael Purdue $400 for expenses and sent them off to Quebec, then the capital of the United Canadas, to lobby for Federal approval on Brampton's behalf. The following year the two men travelled to Quebec again, this time armed with $600. At last the matter was settled in Brampton's favor. A provisional county

J.W. Cole, Brampton's first commercial photographer, opened his studio in the 1850s on Main Street South. Many of his glass negatives, processed over 100 years ago, are still in good condition. Courtesy, Region of Peel Archives

```
FIRST PRIZE
PHOTOGRAPH
AND
FINE ART GALLERY,
(Next door to J. Mathers,)
MAIN STREET, BRAMPTON, C. W.

J. W. COLE, PROPRIETOR.

PHOTOGRAPHIC LIKENESSES
Taken in Every Style of the Art,
FROM ALBUM TO LIFE SIZE.

PORTRAITS
DRAWN ON CANVAS, PAPER, IRON, OR GLASS,

And by the introduction of recent improvements, I am enabled to make
a greater variety of, and more perfect pictures, than have ever been
attempted before in this place.

Stereoscopic Views of Families or Houses
TAKEN TO ORDER.

ALBUM BOOKS AND STEREOSCOPES FOR SALE.

REMEMBER THE PLACE
J. W. COLE,
MAIN STREET, next door to J. Mathers', BRAMPTON, C. W.
```

council was set up in 1865. R.A. Hartley and John Henderson represented Chinguacousy, and Christopher Stork, a pharmacist, spoke for Brampton. In retrospect the decision to make Brampton the county town was deemed to be a good one. Malton was considered to be too far east, Streetsville too far west.

When it came time to choose a site for the new courthouse and jail, those with influence bickered, as they had done with the railway station, over who should donate land. At least a dozen major properties were offered. Several architects bid for the job. At last the county buildings went up. The final bill was $40,000 and in January 1867, the same year as Confederation, they were the scene of the first meeting of the newly-formed Peel Council. Robert Smith, R. Hamilton, and J.P. Hutton represented Chinguacousy. Kenneth Chisholm stood for Brampton. As a concession to Streetsville, which had put up such a vigorous fight to become county town and had become incorporated in the

INDUSTRY COMES TO TOWN

process, Dr. John Barnhart, Streetsville's reeve, was appointed the first Peel County warden.

A small news item in the *Brampton Times* of February 28, 1868, heralded another important stage in Brampton's history—its incorporation as a town. Sandwiched between a tale about a judge being caught in a snowstorm and the proposed impeachment of U.S. President Andrew Johnson was the following snippet: "The Bill for incorporating Brampton as a town has passed its second reading. It is now sure to become law but will not take effect until next January."

Brampton photographer J.W. Cole's studio on Main Street as it looked in 1868. Note the dirt road, wooden sidewalk, and hitching posts for horses. Courtesy, Region of Peel Archives

In fact five years were to pass before incorporation came about. In 1873, under Chapter 51 of the Statutes of Ontario, Royal Assent was given "to Incorporate the Town of Brampton." The first town election was held in January 1874. John Lynch, always a "Brampton Booster" (a term still used today), celebrated by publishing a Peel directory.

The village was ready for incorporation as a town—according to a Dominion census taken in 1871, Brampton had 2,090 inhabitants. Its properties were assessed at $330,440.

As plans for the Grand Trunk Railway (later Canadian National) had preceded Brampton's incorporation as a village, so plans for the Credit Valley Railroad (later Canadian Pacific) preceded Brampton's incorporation as a town. George Laidlaw, who had done much to promote Hurontario Street, did a great deal of leg work, sending pamphlets to municipal bodies and farmers, and travelling and making speeches in the areas through which the railroad

THE COMMUNITY'S POLICE

What started as a one-man operation in the small town of Brampton over 100 years ago has now grown into one of the most formidable law enforcement operations in the country.

Brampton's efforts to combat crime certainly started under modest circumstances in 1873 when a by-law was passed by the newly formed Brampton town council. The by-law called for the induction of John Hurst as chief constable at the princely salary of $40 a year.

By all accounts, Hurst was hardly overworked in the small rural community. His duties, as well as those of his assistant, Stiles Stevens, were little more than to act as night watchman for the small downtown area and check doors to make sure that they were properly locked.

This was the way police work was done in Brampton for some time. Andrew Herkes is a classic example of the way small-town law enforcement worked in Brampton, as well as in the hundreds of other small communities throughout the country. Herkes was a former game warden from Scotland, who had come to Canada in 1912. By 1914 he found himself chief constable of Brampton. The former game warden is still remembered by some in Brampton as a well-liked fellow who was active in the town, especially when it came to sports.

Herkes's popularity with the townspeople put him in an ideal position to keep informed on what was going on around town. He would sit in the police station by the wood stove and people would drop by to visit and tell him what was going on and where. It was community police work at its best.

He was also known for not dragging every first offender to court if he thought that a good talking to would do the trick. So laid-back was the chief that on one occasion when he found it would be necessary to arrest two offenders at gunpoint, he had to go back to the station for his gun.

The policing profession remained a simple one in Brampton throughout the 1940s, 1950s, and midway through the 1960s.

Frank Keats was chief from World War II until 1965. During this time, the police in Brampton were also responsible for the ambulance service. Therefore, if there were many serious injuries at a particular time, the town might be without police while the patients were taken by ambulance to Toronto.

Since radios in police cars were not introduced until the 1950s, prior to that time some way had to be found to contact the constable on the beat when he was needed at the station. Keats solved this problem by working out a system whereby streetlights near the Four Corners area were flashed in the hope that the officer would eventually notice and report back.

The Brampton police were investigated by the newly formed Ontario Police Commission in the mid-1960s, and Chief Keats was criticized for exhibiting bad management skills. Blame could hardly be placed entirely on Keats, for up until that time there was little uniform training for police officers. One of the commission's recommended changes was to relieve the police of their ambulance duties.

The last chief of the Brampton police force was Stanley Raike, who was in charge of the force until the Peel Regional force was established in 1974. Raike was made deputy chief in the Peel Regional Police.

Brampton has changed significantly since Chief Herkes made his rounds, both in size and in economics, and law enforcement techniques have kept pace with the changes. Brampton now joins Mississauga, Chinguacousy, Streetsville, and Port Credit to make up the Peel Regional force, which employs over 800 officers. State-of-the-art communication equipment allows it to be in touch with other law enforcement agencies instantly. As well, contrary to decades past when there was no formal training for new officers, the Ontario Police College is now one of the most respected training facilities in the world. Be it the town force or part of the larger Peel police, law enforcement in Brampton has worked to maintain the highest-level community protection, and there is every indication that the trend will continue.

—Ken Moore

INDUSTRY COMES TO TOWN

After serving as Reeve of Brampton, Kenneth Chisholm was elected Warden of Peel. In 1873, he became the Liberal representative for Peel in the Ontario Legislature. "Alderlea" (left) was his grand house and gardens. Courtesy, Historical Atlas of Peel County

was to run.

The railroad's charter was issued in 1871, but it ran into problems before it was even built. Despite generous donations from the public, materials had to be bought on credit, and rival railways did their best to stop the new company from obtaining traffic rights to Toronto. But in 1879, the new railway was opened by the Marquis of Lorne, the son-in-law of Queen Victoria and Canada's Governor General.

Unfortunately, that wasn't the end of the railroad's troubles. It was poverty stricken right from the start, and a few months after it started operating, the employees went on strike. According to stories popular at the time, the owners were in such bad financial shape that they had to frequently borrow coal from the Grand Trunk Railway. Passengers were apparently kept waiting in Toronto, while a few cartloads of fuel were shovelled into the tender. However, the railroad's troubles were solved when Canadian Pacific took it over.

Meanwhile, Brampton had access to another network of railways, and the outlying hamlets' and villages' isolation became even more obvious. With their water-powered mills and machinery, they were fast becoming obsolete.

A description in the 1877 *Historical Atlas of Peel County* gives a good picture of how Brampton was progressing:

Brampton has kept on increasing in wealth and population until it is now a smart town, with a population according to the last count, of 2,551. The town is growing in wealth from year to year very rapidly, this is perhaps owing in some degree to the number of farmers who have retired, bought property and settled in the place. The total assessed value of the town is $606,575.

By this time Brampton had seven churches—the Church of England, Canada

BRAMPTON

By the end of the nineteenth century, Brampton's network of railroads had linked it with other towns and villages, encouraging its commercial and industrial growth. Outlying villages of the Gore and Chinguacousy, however, suffered from their relative isolation, as demonstrated by the 1925 view (below) of Tullamore Main Street (Airport Road). Timetable, courtesy, City of Brampton; Tullamore, courtesy, Region of Peel Archives

INDUSTRY COMES TO TOWN

Methodist, Primitive Methodist, Methodist Episcopal, United Presbyterian, Free Presbyterian, and Baptist. There were four schools, two banks, two telegraph offices, five good hotels, and several general stores.

The general store was very much a feature of nineteenth-century life. It was, in some ways, a smaller version of our present-day plaza, but instead of going from store to store, one could purchase everything in one place. For example, there might be a dry goods counter, a shoe section, a grocery department, a liquor department, rubber boots, overalls, gloves, nails, tools, pots and pans, stoves, lamps, and coal oil, all under one roof.

Such stores were more than mere markets. They were meeting places where people liked to gossip and catch up on news. They had a friendly, intimate atmosphere and the personal approach was the order of the day. The *Brampton Times* of January 24, 1868, carries an advertisement placed by William Hurst:

William Hurst begs leave to return thanks to his numerous Customers and Friends in Brampton and the County, for their kind support for the past nine years. . . .

This was a preamble to a notice about a sale:

Selling Off!
Dry Goods, Ready-made Clothing
AND BOOTS AND SHOES
He will sell Cheap for Cash "as Cheap as the Cheapest"
until the 1st APRIL,
to make room for Spring Importations
In the GROCERY DEPARTMENT will be found
an Excellent Lot of
TEAS
In the LIQUOR DEPARTMENT Good Wines, Brandies and
Whiskies

Apart from the capital letters and exclamation marks, which were customary in nineteenth-century advertisements, it was a courteous, low-key approach—hard to imagine in today's brash, ad-oriented society.

On the other hand, it is surprising to find that liquor was often sold in such general stores. The current debate in Ontario, about whether liquor should be taken out of the hands of the government-controlled liquor stores and sold in corner shops, is nothing new, it seems. By selling liquor in corner stores we would merely be reverting to what was the norm a century or so ago.

Not everyone approved, of course. John Watson, who had used his considerable influence to sway the vote for Brampton as county town when Malton was in the running, was a strong advocate of abstinence. One of the first things he did when he immigrated from Aberdeen was to form the Total Abstinence Society. Under its rules the use of liquor was prohibited on farms and places of business. Drinking was considered a social evil and an impediment to the main matter at hand, namely, earning a living.

This obsession with temperance came to the forefront several times in

An early creamer manufactured by G. Fulfer in Brampton. By the end of the nineteenth century, the farm business was in decline but other industries in Brampton's diversified economy were springing up to take its place. Courtesy, Region of Peel Archives

Brampton's history. Richard Blain, founder of Blain's Hardware on Main Street North, was a staunch teetotaller. A fervent supporter of the anti-liquor movement, he became a Town Councillor in 1885. Five years later he became Brampton's Conservative member of Parliament. Later he was called to the Senate. Throughout his political career he fought avidly for his cause and he was behind every temperance bill proposed for Peel County.

Businessmen such as Blain who were politicians as well were common in Brampton's early days. With their money and influence, they did much to advance the development of the town. John Haggert, founder of Haggert Bros., became Brampton's first mayor in 1865, after it was incorporated as a town. Allen Franklin Campbell, publisher of *The Conservator,* was chairman of the school board and president of the Peel Agricultural Society. He too served as a mayor of Brampton, from 1887 to 1888.

Brampton's entrepreneurs were also active in Federal and provincial politics. There was strong competition between the Liberals and the Conservatives. George Wright, who brought the first railway to Brampton, was an ardent Orangeman who was also a staunch Conservative. He represented

THE VOICE OF BUSINESS

Boards of Trade have existed in Brampton for over 100 years. The first was founded June 14, 1887, at a meeting in the town's Council Chamber, then located in the old Orange Hall. Kenneth Chisholm was the Board's first president. However, after officers were elected, the organization lapsed.

By 1890, a new Board of Trade was formed. E.O. Runians, a local merchant, was elected president. This new Board helped bring the Young Brothers Foundry to Brampton in 1890 and secured a purchaser for the Haggert Brothers Foundry after it went into liquidation in 1891. But, by 1897, this Board, too, ceased operations.

The Board of Trade was again revived in 1903, and, assisted by the town council and encouraged by *The Conservator*, took steps to ensure that Brampton's new industrial base would not be eroded away, by encouraging new industries to locate here and building an infrastructure of urban services.

The founding of this new Board of Trade marked the beginning of an era of business prosperity, transforming Brampton into a progressive town with hydro-electric power, a dependable water system, sewers, good roads, a new library, and several new industries. The Copeland-Chatterson Company, Hewetson Shoes, Lindners Ltd., Pease Foundries, and Consolidated Crossin Piano Company were among the companies that came to Brampton as a result of the efforts of this small group of businessmen, working hand-in-hand with the town Council.

During the 1920s, the Board of Trade was eclipsed by the Brampton Business Men's Association. Led by local merchants such as R.I. Blain, R.D. Boyle, and W.C. Bartlett, the Association addressed matters affecting the old downtown commercial area, such as persuading the town council to regulate parking and establishing the first off-street parking lot. The Association is best remembered, however, for its annual community picnic in Eldorado Park. After 1931, the Association disbanded due to declining interest.

In 1947, following a proposal by W.T. Rayson of the Copeland-Chatterson Company, the Brampton Chamber of Commerce was inaugurated, chaired by Cecil Carscadden. Officers included President F. Gordon Umphrey and Secretary Archie G. Stephens.

The young organization promoted civic, commercial, industrial, and agricultural progress, co-operating with the town council to attract new industry to Brampton. The council and the parking committee found a solution to the town's perennial parking problem; the three main lots still in use today were laid out. The Chamber's retail section concentrated on improving the downtown area's economic health, resulting in sales forums for merchants, standardization of store hours, and other reforms. By 1960, the Chamber of Commerce had achieved considerable influence despite lagging membership, operating with only a part-time manager, and the fact that it had no office of its own until 1958.

In 1964, in an effort to heighten the Chamber's appeal to the growing Bramalea area, it became known as the Brampton and District Chamber of Commerce. However, by 1975, after a preference had become apparent among members for the term "Board of Trade," the Chamber assumed the name of its old predecessor, the Brampton Board of Trade. Located in its present quarters in the old Carnegie Library, the Board was still a comparatively small group—approximately 300 members—not yet able to reach the large new industries that had moved into Brampton since the late fifties. But, a successful membership drive resulted in an unprecedented membership increase. Today, the Board has 900 members and hopes to reach 1,200 by June 1988.

Currently, under General Manager Nancy Lumb, the Board is improving its services to the business community. It offers a variety of seminars on topics such as management; it sponsors social functions for business people to meet and talk with other business people; and it forms policy so Brampton business can be represented as a unified voice before the various levels of government.

As Brampton grows, the Board of Trade has committed itself to keeping pace with that growth—being the "voice of business for business"—thus doing its part toward keeping Brampton a continually prospering community.

—Naomi Brusse

One of Brampton's most successful early entrepreneurs was John Haggert. His company, Haggert Bros., manufactured farm machinery and for forty years it was Brampton's largest employer. Haggert's home, "Haggertlea" (facing page), was one of Brampton's finest. Courtesy, Region of Peel Archives

York West Riding (of which Peel was then part) in Canada's fourth Parliament, prior to Confederation.

Kenneth Chisholm was a man of very different political hue. After serving as reeve of Brampton, he was elected warden of Peel, and in 1873 he became the Liberal representative for Peel in the Ontario Legislature—a position he held for six terms. A successful merchant and grain dealer, he was a director of the Central Bank of Canada and vice-president of Haggert Bros. He organized Brampton's first Board of Trade. Despite his obvious business acumen, Chisholm sometimes let his political feelings get the better of him. According to some local sources, he was so miffed with the Conservatives for taking the credit for Brampton's railroad that he dismantled his house and turned it around. That way, he reasoned, he wouldn't have to look at an "enemy" railroad from his window.

One of the most successful of Brampton's entrepreneurs/politicians was

INDUSTRY COMES TO TOWN

John Haggert. Haggert, who came from Paisley in Renfrewshire, had, in the best Scottish tradition, apprenticed as an engineer. By the time he was twenty-one he was working on a St. Lawrence steamer.

Haggert came to Brampton in 1849. He opened his farm implement business at just the right time. Farming was backbreaking, badly paid work and manpower was becoming scarce. Families were moving in and out of the community at an astonishing rate. Young men, lured by the high wages being offered by railway companies, left for other parts of Canada and the United States. But as the labour force dwindled, farming became more efficient, partly due to improved methods and partly because it was fast becoming mechanized.

When Haggert Bros. started it employed ten people and produced horse-drawn threshing, reaping, and mowing machines. Two years later steam was introduced and Haggert's turned out some of the finest machinery in the land.

Its products included steam engines and boilers; the Brampton Triple Harvester and Self-rake; the Simple Reaper; the Beaver Mower; Brampton Sulky and Horse Rake; Plaster Sower; the Hay Tedder; feed mills; and the Credit Valley Stove. John Haggert took an enormous pride in their products and he made sure that people knew how to use them. Instructions for a threshing machine in a contemporary catalogue admonished the purchaser:

Don't expect your machine to run without horses, oil and good common sense management. Remember that the man who earns most money with a threshing machine is he who gets up early and keeps it going at a steady jog all day, rather than he who rushed it through for an hour or two, and then loses time in idleness or tinkering.

The machinery was crafted as carefully as a work of art, and according to the *Historical Atlas of Peel County,* "the working with it was almost a pleasure." The following description of Haggert's horsepower grain separator comes from the same source:

These were pronounced by experts who examined them to be the most perfect machines for threshing and cleaning grain which they had ever seen and even to an uninstructed eye it seemed to work like charm. The machine was elegantly gotten up of various Canadian woods and had received an exquisite finish equal to any cabinet work. . . .

Even allowing for the gushy prose that was characteristic of the time, there's no doubt that Haggert's machines were exceptional. The company was at the forefront of technology and its equipment was becoming very popular. The Peel Atlas records that:

The reputation of their manufactures created such a demand, as compelled a steady and continued increase until, at the present time, they occupy the very front rank amongst the agricultural manufacturers of the Dominion, in proof of which they point to the fact of having taken prizes at every provincial and county exhibition for the last fifteen years.

Haggert also won prizes in Sydney and Philadelphia. The *Boston Journal of Commerce* described the horsepower separator as "the most perfect machine for threshing and cleaning grain which they had ever seen." And the company sent its goods to other parts of Canada. A thresher, the first in the province, was shipped to Manitoba. It went via St. Paul, Minneapolis, by train, river boat, and oxen trail, and transportation alone cost $350. Soon Haggert had several customers in Western Canada (later he opened a foundry in Manitoba) because the larger farms in the prairies required larger machinery.

The Brampton foundry looked as impressive as its machinery. The factory sprawled over two acres of land. The main building was four stories high. It was built of brick with stone facings and white brick arches over the window. As befitted a Brampton baron, Haggert also had a fine home. "Haggertlea," as the mansion was called, had grape arbours, a fountain and a large terraced lawn. Situated at Nelson and George Streets, his house has been converted into apartments.

By the late 1870s, Haggert had 140 employees with a payroll of $60,000—an incredibly small amount for that number of employees by today's standards, but then it was regarded as generous. Haggert did so well that he had to give up his municipal duties. After serving as Brampton's mayor for three years, he resigned when he bought the St. Thomas Agricultural Works.

Given the foundry's huge success, it seems astonishing that two decades later Haggert Bros. had gone out of business. After pushing pioneer farmers into the age of technology, Haggert's in turn became technology's victim. With

A drawing of John Haggert's "Haggertlea," made circa 1877, offers a good view of the house and grounds, including the grape arbors (left), fountains, and large lawn, sometimes used for outdoor games. Courtesy, Region of Peel Archives

INDUSTRY COMES TO TOWN

An early steam engine, photographed in 1966. This tractor was built in Waterloo, Ontario, and is very similar to those manufactured by Haggert Bros. in Brampton. The driver was Bill Bailey, the "Oil Man" of Brampton. Courtesy, Guardian Collection, Region of Peel Archives

In 1860, Harry Dale, founder of Brampton's flower industry, opened for business. The Dale Estates achieved worldwide fame. These employees were gathered in the Dale shipping room around 1890. Courtesy, Region of Peel Archives

the introduction of the furnace as a means of heating homes and businesses, Haggert's stoves became obsolete. The bulk of the wheat business had moved to the prairie provinces and the extension of the railways into Western Canada meant that Peel farmers no longer had a monopoly on selling wheat.

Toward the end of the nineteenth century, Brampton's boom had ground to a halt. Farmers had to redirect their enterprises into mixed farming. There was an economic recession. Farms were sold (the *Brampton Times* was full of auction notices at that time), and many farmers moved. Haggert's went into liquidation in 1891, taking with it many of the smaller "satellite" businesses that had relied on its trade.

But the death of one era heralded the birth of another. By 1914 Brampton had ceased to be a farm-oriented service town and had become a centre for industrial manufacturing. The Brampton Knitting Mills (now a restaurant and office complex) were churning out clothes. There were cabinetmakers such as Phillips and Bryant and Henry Burnett. Burnett made everything from bedsteads to cane chairs, and he produced, of all things, "fashionable" hearses.

Dale Nurseries was originally a market garden founded by Edward Dale in 1860. Dale had started his business in a humble way, peddling vegetables from door to door in a wheelbarrow. When his son Harry joined the business ten years later, he persuaded Edward to switch to flowers. By the turn of the century, Dale's was a thriving concern.

Exciting innovations were taking place. Electricity and the telephone arrived. Not surprisingly, these new developments were the cause of many a

The Dominion Building, on Queen Street East, is Brampton's most prominent landmark. Designed by Thomas Fuller in 1889, it was originally built to house the post office and customs and inland revenue offices. Fuller's most famous work was the original block of the Parliament buildings in Ottawa. This is one of the few examples of his work that remains. Courtesy, Brampton Public Library

lively debate. There was quite a controversy over whether or not electric lighting would replace gas. An 1883 editorial in *The Conservator* proclaimed that "electric light has had its day." The writer went on to expound the merits of the gas lighting system, then operated by John Haggert. Unfortunately, he had no way of knowing that not only would Haggert go out of business, but the application of electricity would go way beyond lighting.

A primitive generator was installed on the Credit River in 1885 by James P. Hutton, yet another politician/entrepreneur, to operate his woollen mill. Regarded in local engineering circles as "an important contribution to the new technology," it produced on 100 horsepower. Hutton then built a 2,200-volt line into Brampton to supply power for lighting. An arc lamp was installed in front of the Queen's Hotel, and it shone so brilliantly that "it enabled people as far distant as the Grand Trunk crossing to read the small type of the newspaper."

Hutton's experiment was considered so successful that Brampton promptly ordered six lights for the town's streets. Within months a dozen more were in place. The lighting system cost $1,300 per year to operate, and to conserve power it was shut down at 12:30 every night. It was even switched off during a full moon—a not-uncommon practice in those days.

Despite the success of the street lights it was awhile before electric lighting was used indoors. In the 1890s, arc lights were installed in Christ Church, and although the illumination was good, the noise was so loud that the old gas lamps were brought back into operation.

The Huttonville dam was damaged several times when the Credit River flooded its banks, but electricity continued to be produced. In 1903 John McMurchy bought the dam and expanded and improved the power plant. Not only was it able to supply lighting, it was able to generate power for business—most notably to the Brampton Knitting Mills, the first business to utilize electricity. By then, forty-three customers were benefiting from Huttonville's generating station. Brampton's desire to keep abreast of the times was also evident in its early adoption of the telephone. Residents were given a demonstration of Alexander Graham Bell's new device on September 20, 1877, just a year after he had introduced it to a marvelling public at the Philadelphia Centennial Exposition.

A group of curious residents, many of them downright skeptical, gathered at the telegraph office of P.L. Wood to listen to this new-fangled machine. A series of experiments were carried out between Brampton and Stratford, seventy miles away. The participants talked and sang, and *The Convervator* carried an editorial on this novel event:

Though the atmosphere was not favorable for the experiments at the time of the trial, the words spoken at either end were quite distinct most of the time, and the music of the songs so plain that anyone familiar with the tune would readily catch the melody.

The songs were transmitted so distinctly that to listen to them was a real musical treat. The sound appeared as though come from a short distance away, and whether the tone was low or high it was equally distinct. The telephone is certainly one of the greatest triumphs of the age.

The article captures the curiosity and excitement that the telephone invoked in people at that time, but despite the wide-spread enthusiasm, Brampton didn't get a phone system until the middle of the 1880s. In 1884, the first telephone

This picture of a man with a "penny farthing" bicycle was taken in the studio of J.W. Cole, Brampton's first commercial photographer. Courtesy, Region of Peel Archives

Brampton telephone operators circa 1911, seated at the No. 1 standard magneto switchboard. They are wearing the breast-type transmitters then in use. The local manager was A.L. Cook. Courtesy, Brampton Public Library

exchange was set up in a jewelry store owned by Algernon Williams. He became the first agent for the Bell Telephone Company, and the first telephone directory was issued the same year. Twenty-three subscribers were listed, most of them businessmen.

Activated by a hand crank, the first telephones were bulky, crude instruments with wet cells or "batteries" encased in wooden boxes. The telephone wires were made of iron, and as there were few poles, the wires zig-zagged from building to building. Numbers were not in use in the early days. If you wanted to speak to somebody, you simply asked the operator for that person by name. With such few subscribers, it was easy to track someone down.

One of Brampton's earliest operators was J.S. Beck, who later became a

A Brampton construction crew of the Bell Telephone Company of Canada. Brampton's first telephone directory was issued in 1885 and listed twenty-three subscribers. By 1914 the list had grown to 535. Courtesy, Brampton Public Library

reeve, then a warden of Peel County. He was mayor of Brampton several times. While learning the jewelry and watchmaking business at Williams' store, he doubled as a telephone operator and repairman.

In 1894, E.W. Knowles, a druggist, took over where Algernon Williams left off. He became Brampton's next telephone agent and he was succeeded by W.R. Sharp, another druggist. At that time, telephones could be used throughout the day (prior to that the exchange was operated only from 8 A.M. to 8 P.M.) but Sunday service was restricted. Subscribers could only utilize their phones between 8 and 9 A.M.; 10 A.M. to 2 P.M. and 5:30 to 9 P.M.

Not only was service restricted, communication was often very difficult. People had to shout to make themselves heard. The phone service was a long way from present standards.

Richard Elliott and a customer in front of the Elliott Blacksmith Shop and Garage circa 1905. In the automobile's early days, blacksmiths often repaired cars as well as wagons and horse gear. The horse and buggy gave way to the automobile in Brampton by 1910, following the auto's first appearance there in 1900. Courtesy, Region of Peel Archives

As the twentieth century approached, however, the equipment and service improved. A copper line was installed between Brampton and Toronto. Previously only a single wire linked Brampton to Streetsville. To drum up business, the Bell Telephone Company embarked on a vigorous sales campaign. A directory issued in November 1902 carried an interesting slogan:

JUST TALK - DON'T TRAVEL OR WRITE

The sales drive and the improved service boosted the number of subscribers. By 1905 the telephone had become a necessity, and Brampton had 100 subscribers.

The horse and buggy era was fast drawing to a close and with it went several industries. The automobile had arrived. It made its first appearance in Brampton in 1900, when Lord and Lady Minto visited the Dale greenhouses. Not long after their visit, W. Emerson Downs and his father put together a "horseless carriage" at their machine shop on Queen Street East. Soon garages and service stations put saddle and harness makers out of business. Tailors, coopers, and tanners disappeared. In their place came salesmen, electricians, architects and accountants. Another phase of Brampton's history was about to begin.

This is the Stork family, photographed in Eldorado Park. Jimmy Cooper, a friend of the family, is seated beside the driver. This was one of the first Ford autos in the area. Courtesy, Region of Peel Archives

CHAPTER FOUR
SPIRIT, MIND, AND VOICE

The backbone of any community is its social structure—religious outlets, communications system, sports and leisure facilities, and civic services like hospitals and fire protection. Although very basic when compared to the wide range available today, Brampton had already laid the groundwork for many of these organizations during the times of the pioneers.

By the 1850s, the pillars of a Victorian society were in place. There were churches, schools, doctors, newspapers, a library, and various sporting organizations. During the late 1800s and early 1900s, Brampton moved from an agrarian society into an industrial and technological one. Organizations and services became more widespread and more sophisticated. The foundation of Brampton's social structure as we know it today was laid by its early citizens, who left a legacy from which Bramptonians are still benefiting.

In the early days, entertainment and leisure were luxuries. In an era where people were concerned with working all day merely to put food on the table and a shelter over their heads, there wasn't much time left over. But even as early as the 1820s "raising bees" had become part of pioneer life. Groups of settlers got together to help one another build farms, barns and houses. This practice came about because of a shortage of manpower, but it was also a chance for people to socialize and catch up with the latest news—a welcome break in an otherwise tedious and backbreaking existence.

Parades of the militia, held on special days such as the monarch's birthday, were always great sources of entertainment. Ill-trained soldiers performed military drills with sometimes ludicrous results. Afterwards, they staged horse races and other sporting activities. It was common practice to partake liberally of alcohol, and by the end of the day several fights had usually broken out.

An amusing account is given of such an event in Edwin Guillet's *Early Life in Upper Canada*. In 1837, a Mrs. Jameson described a parade-day at Springfield (Mississauga), then part of the County of Peel. She wrote:

The motley troops, about three or four hundred men, were marshalled—no, not marshalled, but scattered in a far more picturesque fashion hither and thither.

A 1950 kiddie's parade on Main Street looking north from Queen Street. Courtesy, Brampton Public Library

A few men, well mounted and dressed as lancers, in uniforms which were, however, anything but uniform, flourished backwards on the green sward, to the manifest peril of the spectators; themselves and their horses equally wild, disorderly, spirited, undisciplined: but this was perfection compared to the infantry. Here there was no uniformity attempted of dress, of appearance, of movement; a few had coats nor jackets but appeared in their shirtsleeves, white or checked, or clean or dirty in edifying variety.

Some had firelocks; some had old swords, suspended in belts, or stuck in their waistbands; but the greater number shouldered sticks and umbrellas. Mrs. M. told us that on a former parade-day she had heard the word of command given thus—'Gentlemen with the umbrellas, take ground to the right! Gentlemen with the walking-sticks take ground to the left.'

Mrs. Jameson goes on to describe how these "soldiers" elbowed and kicked one another or drank and chatted, completely oblivious to the commands of their leader. "Not to laugh was impossible. Charles M. made himself hoarse with shouting out orders which no one obeyed, except perhaps two or three men at the front. . . ."

Other contemporary accounts of holiday celebrations show that such shenanigans were expected. In fact, authorities were surprised if a day passed without incident. Elections, too, were a source of celebration, or more commonly, animosity. Men's political feelings were easily roused when life was hard. The situation was aggravated by the fact that dishonest electoral practices such as open voting, intimidation, and bribery, were common. Elections usually lasted a week, and to make matters worse, they were held in taverns where the candidates dished out booze among the patrons, in order to win votes. Not surprisingly, this practice just inflamed an already volatile situation, but interestingly, this is why in Ontario today it is still the custom for bars and taverns to close during election day.

Brampton's fair was also the scene of many a celebration—as indeed it is today. It started in a small way when market fairs were held at Salisbury's tavern. After the foundation of the County of Peel Agricultural Society in 1853, fairs were held in Brampton, at the corner of Queen and Main. Later it shifted to four acres of land at the corner of Main and Wellington, leased from John Elliott. (Later, this land was sold to accommodate the County Jail and Court.) In 1871 the fair moved again, this time to the corner of Wellington and Mary streets. By 1884 it was becoming obvious that the fair would need even bigger grounds, and thirteen acres of land were purchased at a cost of $3,800, from William McConnell, owner of the Brampton Driving Park.

The original purpose of the fair—apart from providing entertainment—was to give farmers an opportunity to exhibit their livestock, and to buy and sell produce. This is still the case today, although there are all kinds of other activities as well—from a midway to an antique car show. Brampton Fair is a testing ground for many of Peel Region's prize animals—particularly horses and Holsteins—and farmers who have won awards at Brampton go on to do well at major exhibitions, such as Toronto's C.N.E.

(Canadian National Exhibition) and the Royal Agricultural Winter Fair, also held in Toronto.

Today modern technology is on hand to aid the farmer, but in the early days the Brampton Fair showcased heavy draft horses. Strong and placid, such animals were the major source of power. Prior to 1900, the breeding and selling of draft horses was considered more profitable than cattle. By the 1920s, tractors had started to replace them, but horses are still the highlight of the Brampton Fair. Many classes of "pleasure" horses are shown. In fact, horses have made a comeback, so much so that the Brampton area has more today than it did in pioneer times.

Apart from the annual fair, there were many other types of entertainment in the nineteenth century. Travelling circuses and menageries (often the animals were stuffed!) were in great demand, as were travelling theatre companies. Theatre troupes went from town to town, presenting plays and amusing skits, and their arrival was heralded by colorful street parades. The spectacles they presented were often repetitive and predictable. The audiences knew exactly what was coming next. But that was part of the fun; people enjoyed being able to follow the action.

According to the *History of Peel County,* published in 1967, Brampton's particular favorite was the "Uncle Tom's Cabin parade":

A calliope playing Dixie songs was in the lead. Then the venerable Uncle Tom with a head of curly white hair and Little Eva in a pink spangled gown. In the next wagon would be Simon Legree with fierce looking mustache,

Jack Houck of Brampton accepted from M.B. Nichols the prize for the grand champion Holstein at the 1966 Brampton Fair. On the right is Bert Stewart, who showed the cow. Courtesy, Guardian *Collection, Region of Peel Archives*

BRAMPTON

Byrnell Wylie, from Oakwood Farm, demonstrates his ploughing skills. In the 1800s, the Brampton Fair showcased heavy draft horses. By the 1920s, tractors had begun to replace them, but horses are still one of the highlights of the Brampton Fair. Courtesy, Guardian *Collection, Region of Peel Archives*

SPIRIT, MIND, AND VOICE

wielding his long cat-o-nine tails. Liza brought up the rear with her baby wrapped in a shawl. Finally came the blood hounds, barking and straining at the leash, held by a strong negro lad. Could anyone stay away after seeing such a graphic pre-view?

To drum up business, professional entertainers advertised in the local press.

In return for a fee, "elocutionists" would attend church gatherings and garden parties, and their repertoire ranged from Shakespeare's soliloquies to "The Minister Comes to Tea." Bramptonian Lily Mae Kee was an excellent mimic who apparently could imitate an Irish brogue to perfection. Her humorous sketches were said to bring tears to the eyes of her audience.

Such simple and mostly "homemade" entertainment was the order of the day, but as the nineteenth century drew to a close, it became more sophisticated. Live bands were big. Waltzes, reels, and square dances were all the rage, particularly in rural areas. Castlemore (now part of the City of Brampton) was famous for its lively Friday night dances. A group of young men called "The Young Bachelors" hosted get togethers every other Friday evening and couples came from as far afield as Alliston and Tottenham.

In *From This Year Hence*, a history of the Toronto Gore, George Tavender writes:

It was quite permissable for a young man to escort several ladies to these affairs, which also included a late snack for the sum of twenty-five cents. Mr. O'Leary operated the wagon shop at this time and used the hall as a paint shop during the week, but on the night of the dance, the hall would be spic and span and gaily bedecked with bunting and evergreens.

Churches also provided entertainment—although, to be sure, of a rather genteel nature. Church groups organized oyster suppers, strawberry festivals, garden parties, and chicken suppers. Sunday School picnics were very popular. The church, of course, was the mainstay of family and spiritual life. It had been so since Brampton's founding.

But the history of Brampton's religious life started even before the arrival of its founders. John Elliott and William Lawson, Primitive Methodists, were probably the first people to organize congregations, but prior to that, Brampton was serviced by circuit riders—travelling ministers and priests.

It was the custom for itinerant preachers to wander the county, and they spent months going from place to place. Zealous men intent on bringing God's word (and incidentally, news) to the hinterland, they travelled on horseback, at times hacking their way through dense bush, sometimes traversing bumpy corduroy roads. Sometimes preachers lost their way in the heavy bush. An interesting story is told about a Methodist preacher who was travelling in Chinguacousy Township. On the appointed day, people gathered to await his arrival at a designated spot in a clearing. They waited till midnight but he still didn't show up. Undaunted, the congregation bedded down for the night, curling up in the blankets on the ground. Suddenly, through the still night air, came the distant sound of a hymn. The preacher was approaching. The settlers sprang into action. Bellowing out the choruses

Marlene Eccles, of Mayfield High School, demonstrates her farming skills at Brampton's annual fair in 1973. As in pioneer times, the fair continues to attract people of all ages. Courtesy, Guardian *Collection, Region of Peel Archives*

The Brampton Mechanic's Band, later known as the Brampton Citizen's Band, is believed to have been founded in the latter part of the nineteenth century, when this picture was taken. In 1922 it won its first musical competition at the Canadian National Exhibition Band Contest. Courtesy, Region of Peel Archives

they knew and loved so well, they piled logs on the fire. Attracted by the noise, the preacher eventually found them, and the service went on, under a sky filled with stars.

Because the roads were so bad and the areas they covered were so vast, circuit riders could only visit their congregations once every six weeks or so. They were assigned various appointments on a rotating preachers' plan—set up in much the same way as a school timetable. Between visits, deputy ministers—laymen of various ranks—ministered to settlers' spiritual needs.

The settlers, cut off from the outside world, looked forward to the visits of the circuit riders. Many were willing to travel for miles to hear a preacher speak, even in the depths of winter. They thought nothing of trudging through forests with snowshoes or heavy sledges. Ministers and priests were often the only people in the community with a college education. They were much respected for that reason and their role went beyond that of preacher. They were amateur psychologists as well. To people whose lives were dogged with illness and filled with days of drudgery, the circuit riders brought hope and understanding.

Initially, religious gatherings were held in the parlors of taverns or private homes. As communities developed, simple churches were constructed, like the houses, from logs. Congregations with permanent ministers came into being later, when communities expanded and stabilized. There was a mind-boggling array of sects. In the mid-1800s, there were many branches of Presbyterians and at least five different kinds of Methodists. There was

SPIRIT, MIND, AND VOICE

The Reverend Garnet Watson Lynd was one of the late-nineteenth century preachers who rode from place to place to minister to his flock. Such preachers often spent months travelling around the country attending to the spiritual and social welfare of their congregations. Courtesy, Port Credit Library

strong rivalry between these related groups, and in fact, animosities often ran deeper than between people of differing faiths. Even Catholics and Protestants, who had come from widely divergent backgrounds in Scotland, England, and Ireland, tried to get along. In a pioneer society, helping one's fellow man was a tenet of survival.

As the churches became more established they played an important role in the development of Brampton and the other townships. Brampton's first formal congregation was a group of Wesleyan Methodists, who gathered in a private house on Main Street in 1822. From there they moved to the Brampton School House, and in 1848 they erected a building on Elizabeth Street North, not far from Haggertlea, John Haggert's mansion.

In 1866, the Methodists purchased the Scott lot on Main Street North, and built a handsome, brick church, forty feet by sixty feet. It even had a

St. Paul's United Church was built in 1886. A reporter from the Christian Guardian *attended the opening on February 6 and wrote, "This is one of the most complete and beautiful churces in the Dominion. . . . " Courtesy, Region of Peel Archives*

steeple—an expensive luxury in the early days. Grace United Church, as it is called today, was opened in 1867, the year of Confederation. It still stands today on 159 Main Street North, and although it has been greatly enlarged, it still incorporates part of the original building. The sanctuary is believed to be the oldest remaining church structure in the city.

St. Paul's United Church is another Brampton church that can trace its beginnings to the early days. When the Primitive Methodists first arrived from England, they met in the home of John Elliott (site of the Ward Funeral Home). Around 1830, the parent church in England sent a minister to Toronto to organize the York Mission. Eight years later the York Mission was divided into two circuits: York and Brampton. Membership rolls for 1848 show the Toronto circuit had 204 members. For a fledgling settlement, Brampton wasn't doing badly—it had 163 members.

Once William Lawson had arrived, the church was put on a more permanent footing. In 1854, the Brampton church staged the first Conference of the newly independent Primitive Methodist Church of Canada. Towards the end of the nineteenth century, the different factions of the Methodist Church came together, and it was to accommodate the enlarged congregation that St. Paul's United Church was built. The cornerstone was laid on June 6, 1885, by William Gage (later Sir William Gage), a self-made, wealthy publisher who was educated in Brampton. Famous for his philanthropy, he donated Gage Park to the citizens of the town.

The church was completed in 1886 at a total cost of $23,000. A reporter from the *Christian Guardian* covered the opening on February 6. He wrote about the church in glowing terms: "it is not too much to say that this is one of the most complete and beautiful churches in the Dominion. . . ."

Its pale pink exterior was particularly pleasing, but in later years, the exterior became hidden—years of dirt and fumes had turned it a muddy brown. The stonework was cleaned and returned to its original sparkling condition in 1979.

Several other Brampton churches began in the nineteenth century. St. Mary's Roman Catholic Church grew out of the Guardian Angel's Church, which was destroyed by fire in 1878. It had been built on a piece of land donated by John Lynch, two blocks south of Queen Street East. Before the church was constructed, Lynch, like John Elliott and William Lawson, held services at his home. Mass was conducted by priests from the Wildfield settlement.

Christ Church Anglican, the First Baptist Church, and St. Andrew's Presbyterian were founded in the early- to late-1800s. The Salvation Army arrived in 1884. Captain Minnie Leidy set up her headquarters in the market square on which was later built the Carnegie Library. Today the Brampton Board of Trade offices are housed in the original Carnegie Library building.

In the 1820s and 1830s, the function of the churches went beyond providing spiritual guidance. Ministers were often teachers as well, and churches doubled as schools—at least until permanent schools were built. The early schoolhouses were, like other buildings of the time, generally made from logs. As communities prospered, bigger and more solid structures were

SPIRIT, MIND, AND VOICE

Brampton's Methodist Church, built in 1867, still stands today, incorporated as part of this larger structure. Courtesy, Guardian Collection, Region of Peel Archives

St. Andrew's Presbyterian Church, circa 1890. This church was built in 1881 and replaced an earlier building. Photo by J.W. Cole. Courtesy, Region of Peel Archives

erected. The *History of Huttonville School* tells us: "By 1850 it followed that a larger school was needed and a second school was built on the north corner of the Fifth Line, on land purchased from Mr. Wm. Ostrander. This new school was a brick building, heated by a small box stove and lacked such things as slate blackboards, bells or libraries."

Right from the start the pioneers were concerned with educating their children. To parents caught in a web of manual labor, it was important that their children should "get on" and perhaps leave the pioneer life behind, or at least be in a position to improve their lot, as the parents themselves were so often unable to do. As early as 1816 a Common School Law had been passed which ensured that children were taught reading, writing, and arithmetic. In addition to the "three Rs," girls were taught "knitting, sewing, spinning . . . and other useful arts of a like description."

Education was largely the responsibility of private individuals (school boards came later) and Brampton's earliest school appears to have been run by Dame Wright, a kindly but strict lady in the English schoolmarm mould. Her school seems to have catered to very young children. Some sources say that she taught them at her home. Others maintain that she utilized a one-storey building, where *The Conservator* office was later built.

At any rate, various documents prove that she and her husband, a member of the legislature, taught with the best interests of the children in

Brampton photographer J.W. Cole photographed this woman in the uniform of the Salvation Army circa 1885. The Salvation Army had come to Brampton the previous year. Courtesy, Region of Peel Archives

mind, and her school, basic though it was, filled the educational needs of the time. Education was not compulsory—it was not mandated by legislation.

Things changed radically when Egerton Ryerson came along. An energetic and erudite man, he revolutionized education in Ontario. In 1844 he was appointed Upper Canada's superintendent for education, but for years he had been battling to get education out of the hands of churches and private individuals not qualified to teach. Although religious himself (he was at one time editor of the prestigious *Christian Guardian*), he firmly believed that church and state should be kept completely separate. He wanted to replace the haphazard system of schooling with a formal, professional one. Above all, he wanted teachers to be qualified.

One of his first moves was to formulate the School Act of 1850. This act

BRAMPTON

Above: Two unidentified Brampton clergymen. Churches played an important role in Brampton's early development. The local clergyman brought news of the outside world, and often acted as a social director and amateur psychologist as well. Courtesy, Cole Collection, Region of Peel Archives

stipulated that each city, town, or village was entitled to a "school section." Two trustees were to be elected in each ward and part of their responsibilities was to enforce the collection of fees.

Seeking support for these moves, Ryerson, in an impassioned speech to a gathering of newly-appointed trustees, exclaimed: "You are placed in a position to do more for the rising generation of your community than any other class of men. I entreat you to spare neither labour nor expense to establish good schools. It is the best legacy you can leave to those who will succeed you."

Ryerson was right. His school system did indeed leave a legacy to Brampton.

The *History of the Brampton Public School Board* relates that after the act was passed, teachers for the first time had to be "fit and proper men, qualified for the duties of office and ready to fulfil with faithfulness and public spirit the sacred trust committed to them."

The year 1856 saw the construction of the Central Grammar and Public School. Situated on Alexander Street on land bought from John Scott, the two-storey building housed senior pupils on the ground floor, junior pupils above. Adam Morton was the principal, and under his guidance, students were taught grammar, geography, and British history—in addition to reading, writing, and arithmetic. British history was always part of the the public schools—or at least it wanted to give people that impression. Its curriculum in the nineteenth century. Even after Confederation, Canada was

SPIRIT, MIND, AND VOICE

Facing page, right: The old Brampton High School drew pupils from all over Chinguacousy. Opened in the 1800s, its first principal was a Mr. Thompson, who was believed to have been from Ulster, Ireland. This building was destroyed by fire in 1917 and a new one was built to replace it. Courtesy, Region of Peel Archives

Above: These pupils attended Public School in Castlemore, Toronto Gore, circa 1932. Courtesy, Region of Peel Archives

The 1891 graduating class of the Brampton High School. These men went on to work in the local school system, area colleges, churches, or publications. They are (front, l-r): R. Lees, E. Johnston, Alexander Murray, W.J. Galbraith, M. Pilkey; (second row) B. Ledlow, Lawson Caesar, —Lougheed, Ken Peaker, W.A. Kirkwood; (rear) J.C. Kirkwood. Courtesy, City of Brampton

still thought of as a colony, and as such was considered to have no history of its own!

The latter half of the nineteenth century brought tremendous advances in Brampton's education system. As immigrants poured in, the demand for schools increased. An amusing ditty published in the *Brampton Times* reflects this fact:

Facilities to educate
Your children in this Town are great;
If they're allowed to grow up fools,
It will not be for want of schools.

By 1873, 494 students were attending Brampton public schools. Fifty were in the Grammar School. Teaching, although now a recognized profession, was a very badly paid one. Instructors were lucky to receive $300 to $400 per year, which, even then, was a pittance. Education, however, was free—a novel concept in the nineteenth century. John Lynch wrote in the *Directory of the County of Peel* that:

The people of Brampton are entitled to much credit for the liberal support they have given to the cause of education. Since its first establishment as a separate municipality in 1853, the public schools of Brampton have been

A classroom portrait at Brampton Central Public School circa 1921. Courtesy, Region of Peel Archives

perfectly free. Teachers' salaries and all other expenses are paid by the ratepayers of the village, except the portion received from the Government.

He also noted rather snippily:

As the Brampton schools are free to all, there are many scholars from the neighbouring townships who attend the Brampton schools and are thus educated at the expense of the ratepayers of Brampton.

But for those who wanted their children to get the best education going, Brampton was the natural choice. Chingaucousy residents may not at that time have been Bramptonians by birth, but by attending a Brampton school they became, as one resident put it, "Bramptonians in spirit."

The private schools were still in existence but they charged a fee for their services. The 1861 curriculum of the Eclectic Female Institute, a "First Class Ladies' School," shows that board and tuition "per year of forty-two weeks" cost $160. This private school seemed to take its role as seriously as *Course of Study* contains a stern notice "To Parents and Guardians":

We do not regard education as the process of merely adorning the body and storing the mind with information, but the cultivation and development of all the powers of the body, mind and soul. Its aims may be defined to be

Brampton's showpiece Carnegie Library opened in 1907. Today the building houses the Brampton Board of Trade. Courtesy, Region of Peel Archives

gracefulness, systematic thought and correct moral action. No royal road leads to those. The way is a somewhat long one and lies through study, through discipline and through hardships. . . .

An interest in books went hand in hand with learning and when Ryerson drew up his plans, he made provisions for the establishment of libraries as well as for schools. After legislation was passed in 1850, Brampton got its first major library. Opened in 1858, it was run by the Mechanics' Institute. John Haggert, owner of Haggert Bros., was instrumental in getting it going. The book collection was small—it filled only a couple of shelves—but it was eclectic, ranging from Dante to Robert Burns.

Initially, the library had an unstable existence and it moved from place to place until permanent premises could be found. In 1887, the library was

SPIRIT, MIND, AND VOICE

> our name has changed . . .
>
> **The Brampton Times**
> 1855
>
> **The Conservator**
> 1874
>
> **The Banner & Times**
> 1902
>
> **The Times & Conservator**
> 1938
>
> **The Daily Times**
> THE HOME NEWSPAPER OF BRAMPTON and BRAMALEA
> 1973
>
> but . . .
> Our dedication to publish a newspaper which fulfills our community's need to be informed thoroughly and well, remains constant.
>
> **THE PULSEBEAT OF PEEL COUNTY FOR 118 YEARS**

The publishing business in Brampton has been a thriving one since the mid-1800s. This 1973 advertisement shows the various names that Brampton's oldest newspaper has had. Through the years the newspaper has absorbed other publications. Today's Times *is owned by the Thomson organization. Courtesy, City of Brampton*

set up in the Golding Block on Queen Street West. A membership fee of one dollar was established, and grants—$400 from the town council and $160 from the provincial government—put it on more stable footing.

By the turn of the century reading had become more popular largely because people were better educated. It soon became obvious that Brampton needed a better library. Andrew Carnegie, the Scottish-American millionaire, was approached for funds. Carnegie, a self-made man with very little formal education, was well known for his desire to bring learning to the masses. During his lifetime he donated more than 350 million dollars to various cultural institutions. He funded 1,700 public libraries.

Although he was generous, he did not give away money without first making sure that certain standards were met. Brampton initially submitted an elaborate plan, which included a concert hall as well as a library. Carnegie refused to underwrite the concert hall, saying that he would donate $10,000 for a library alone. Eventually the plan was modified and after a meeting with Mr. R.J. Copeland of the Board of Trade, Carnegie provided another $2,500. Brampton's showpiece library opened in 1907.

The love of reading was the natural outcome of a better educated society, but even when education was in its infancy, Brampton had its own newspaper—*The Mercury,* a weekly established by a Reformer named Judd. In 1855 it was bought by the *Brampton Times* (later *The Conservator,* today the *Daily Times*). Brampton had a second newspaper, the *Weekly Standard.* It was later absorbed by the *Peel Banner and General Advertiser.*

Nineteenth-century newspapers were strongly political and reflected not only the viewpoints of readers but also the biases of their publishers. Power struggles between the various publications were common. The editorials today make amusing reading but at the time they were written, they raised political tempers, rousing passions and fuelling the animosities that existed between the many differing factions.

Even events such as an election of village councillors were reported in such a manner as to make certain councillors appear foolish, with no thought of objective reporting or judicious editing. Insults were flung in every direction and were duly reported as if they were of great importance.

The following interchanges recorded in the *Brampton Times* on January 10, 1855, are fairly typical:

On Monday, the 1st instant, the Annual Municipal Election, for Brampton, was opened at the Town Hall. George Wright Esq., was the first to speak. He said, he had always been anxious to serve the people of Brampton, and that he had done much for the benefit of the village; but he had not been able to accomplish as much good for the poor people of Brampton as he had wished, in consequence of the majority of Council opposing every improvement which he had proposed. . . .

The returning officer, John Lynch, replied tartly that this was news to him and that Council had done no such thing.

Later on at the same meeting there was another spat, this time between Wright and a Mr. Ballantyne, another member of Council.

Mr. Ballantyne spoke at some length, and defended the late Council, particularly respecting the erection of the Market buildings, which besides other accommodation, would enable the Village to use the present Town Hall for an additional School-House, which was now much required.

Here he was interrupted by Mr. Wright, who said, that he paid fifty dollars for building the Town-Hall and they should not use that building unless they paid him back his money.

Mr. Ballantyne said he thought Mr. Wright might get back his money, but he did not think any other subscribers would require it.

Mr. Wright said Mr. Ballantyne had not given a shilling towards the building.

Mr. Ballantyne replied, that at the time the Town-Hall was built by subscription, he was an apprentice boy and had scarcely a shilling to give if he had been asked for it, which he was not. He did not think it very dignified in Mr. Wright to taunt him with his poverty. . . .

And so on.

In 1858, George Tye, the composing room foreman of the *Toronto Globe* (now the *Globe & Mail*) bought the *Brampton Times* from Thomas Seller. Tye made the *Times* independent, and when the idea of Confederation first came to the fore, he began to support the Conservative Macdonald-Cartier administration. Annoyed at what he felt was Tye's "defection," Alexander Dick, a Liberal party stalwart from Woodstock, started the *Peel Banner and General Advertiser*. When Dick became the registrar of Peel County, his son Frew took over. He bought the *Times* and later amalgamated the two papers.

A Conservative newspaper—the *Brampton Progress*—was launched by two young printers, Weidman and Hart, three years before Brampton was incorporated as a town. It didn't last, however, and it was left to A.F. Campbell to start another Tory newspaper, aptly named *The Conservator*. Campbell, who spoke so rapidly he was nicknamed "Gatling-Gun," became one of Brampton's leading citizens. He was chairman of the school board and a member of both the Brampton and Peel County councils.

In 1890 Campbell sold *The Conservator* to Sam Charters of the *Woodstock Sentinal-Review*. Charters, a native of Chinguacousy who had served his apprenticeship as a printer in Brampton, founded a publishing dynasty. His sons and grandson continued to run Brampton's newspapers until 1953.

Some of the most interesting reading in Brampton's early newspapers, apart from political diatribes, were the numerous advertisements for "health cures" and "miraculous" lotions and potions. In an age where ignorance about medicine prevailed, people were willing, it seemed, to try anything that would help them feel better. Herb healers and amateur surgeons abounded. Dr. William Johnston, Brampton's first doctor, became famous for his special plaster, which was used, apparently, to cure skin cancer. When made into paste it had "a terrific drawing power." Less publicized was the fact that it also caused intense pain.

Johnston was one of the first settlers in Chinguacousy Township. His log shanty, built in 1818, doubled as an office and surgery. Like many pioneers, Johnston could turn his hand to almost anything. As the population grew, he took on the job of coroner for Peel and York counties. He also taught school, acted as township clerk, conducted auctions, and became Brampton's first postmaster.

But even a man as versatile as Dr. Johnston couldn't cope with the numerous deadly diseases then found in the new settlements. In the 1830s, there was an outbreak of cholera. Twenty years later, navvies working on the Grand Trunk Railway brought another bout to Brampton. Filth and accumulated garbage led to outbreaks of diphtheria. Typhoid and scarlet fever were rampant—as was consumption.

Whole families were swept away by these dreadful scourges. Samuel Book, of the British Arms Hotel, lost his wife and two children to cholera. In the smaller settlements, where transportation was slow and medical care was less readily available, the situation was even worse. In *From This Year Hence,* George Tavender describes how a Scottish family living in the Gore lost all of their eight children. They were buried side by side on their farm, interred in rough wooden boxes that were hurriedly assembled by sympathetic neighbors who shared the parents' grief.

But not everybody was willing to help a family in distress. Not surprisingly, many people were terrified that they too would fall prey to some horrible illness. Tavender relates how when the children of John and Rebecca Lindsay of Tullamore contracted scarlet fever, the neighbours shunned the family. Only a relative and Mrs. Isaac Wilson (Jane Woodill), a minister, came to their aid. Six children died, two of them on the same day.

Wm. James Fenton, interviewed over fifty years ago when he was seventy-three, gave a very clear picture of what life was like in the mid-1800s. As he pointed out, those of us who live in the present tend to see the past through rose-colored glasses, and it's hard to imagine what these early settlers actually went through. Fenton said:

It is characteristic of old age generally to speak of the 'good old days' and recall the glories of the past and simple life, but one who has lived them can only regard them as days to which none of us would wish to return. To be sure taxes were low, but so were wages and the open well at the back door was a breeding place of typhoid. . . .

Fenton was very much a product of his time. Pragmatic and hardworking, he was typical of the men who helped to build the Brampton of today. Orphaned at the age of ten, he was raised on a farm and, as was common at that time, had to work for several hours before leaving for school. He walked to Brampton High School, four miles from his home, every day. Later he graduated from the Model School for teacher training. From there he went on to the University of Toronto. He came back to Brampton eventually, where he was made principal of the High School.

When Fenton was a child, nobody knew much about medicine, and the methods of treatment that did exist were woefully ineffective. New immigrants, for example, were told that in order to avoid getting cholera, they should consume "gentle medicine" such as salts in peppermint infusion, or rhubarb and magnesia. To protect themselves against smallpox, they were instructed to put sulphur in their boots.

Toward the end of the nineteenth century the situation improved somewhat. Brampton's first Board of Health was established in 1864, and regulations were drawn up to counteract malignant and infectious diseases. By the late 1800s, graduates of Brampton High School were going on to medical college, and armed with up-to-date knowledge, they returned to Brampton to practise. Dr. David Heggie founded a strong family tradition when he began practising in 1865, the year Brampton became the county town. He was the father of four sons—three went on to become doctors. Dr. Heggie was a learned scientist as well as a doctor, and many of his medical treatises were published in prominent journals.

In 1884 more improvements were made to Brampton's health system. The Board of Health was revamped, and under Dr. John Turner Mullin (who later became mayor), the board became responsible for monitoring health conditions, and a sanitary inspector was appointed. The inspector visited 600 houses in 1887, and he served notice on those that didn't come up to standard.

Later that year, Dr. Mullin announced that the board had done its best "to effect a thorough sanitary state of the town." Evidently it had done a good job. Only eight cases of typhoid (with two deaths) were reported that year. There were six cases of diphtheria, but none were fatal. As the century drew to a close, great strides were made in the medical field. Doctors were learning new ways of dealing with disease, and surgery had taken a great leap forward thanks to the Crimean and Franco-Prussian Wars. Inoculation against smallpox was introduced, and in Brampton, compulsory vaccination of schoolchildren began in 1895. That year, 535 children were given protection.

As people became more healthy and consequently began to live longer, they turned their thoughts to sports and other leisure activities. Participating in sports was a way of letting off steam after a hard day's work. At first teams from neighboring villages competed with one another. Later they formed leagues and held exchange matches with players from other parts of the province.

Day-to-day life had become less of a grind, and folks found time to enjoy themselves. It isn't surprising to learn that "imported" British games such as soccer, bowling, and curling were very popular. Many Bramptonians, after all, had come from Britain or were descended from British families. But Canadians also invented several games of their own—notably basketball and ice hockey.

Dr. James Naismith, a native of Almonte, invented basketball in 1891. An outstanding athlete at Montreal's McGill University, Naismith was a staff member of the International YMCA Training School at Springfield, Massachusetts, when he was asked to create a game that would fill the gap between the football and baseball season. Now a world-wide sport, basketball is still played by Dr. Naismith's thirteen basic rules.

Ice hockey is believed to have originated in Kingston. On Christmas Day, 1885, men of the Royal Canadian Rifles, looking for something to do, tied on their skates and played with field hockey sticks and a lacrosse ball. (Some historians believe the game was invented earlier). The Ontario Hockey Association (OHA), the first body to govern play and set the rules for provincial championships, was formed in 1891. The following year, F.A. Parker, manager of the Merchants' Bank, organized a Brampton team. The Brampton Intermediates, as they were called, won the OHA's King Edward Cup in 1911.

Many sports groups started in the schools (later they evolved into fully-fledged community teams), such as with soccer, which started with the

BRAMPTON'S ARTISTIC HERITAGE

Perhaps the most famous artist Brampton has produced is Caroline Wilkinson Armington. During her lifetime she was the recipient of a vast amount of praise from the international art world. While living in France, Armington was called "the best woman etcher" in France. A newspaper of her day described her as "an exquisite landscape painter" in praising this other form of her art.

Armington was born in Brampton about 1879. As a young girl she showed an aptitude for sketching. While she was in her teens, she set her life's goal at being able to study art in Paris. In an attempt to realize this goal, Armington moved to Halifax in order to study in French schools. To earn money she taught art in Halifax and later was a nurse-in-training at Guelph General Hospital.

Fortunately, the young girl's wish came true, and she found herself in Paris studying art under F. Schommer, Henri Royer, and Paul Cervais. This period in France accounted for the greater part of her artistic training, save for the time spent in Toronto studying under portraitist J.W.L. Forster.

Armington made several trips around Europe to expand her talent and inspiration. In 1900, she returned to her homeland to teach art in Winnipeg. However, the vast, open spaces and the ruggedness of the Canadian West must have been quite a shock after the metropolitan life of Paris, and she returned to her beloved France in 1905. Although she would return to Brampton to visit her mother on a regular basis, Paris had become her home. From 1908 on, Armington's works were displayed throughout North America, France, and England. She died in 1939.

Norman Price was also an internationally respected Brampton artist and contemporary of Armington. Price favored historical subjects as the focus of his work, and like Armington, went to that mecca of art, Paris, to perfect his skill. After spending time in France and England, Price returned to Canada and became a member of the Toronto Art Students' League. During the early 1900s, he was considered one of New York's outstanding illustrators.

Ronald Bloore, perhaps one of the more outspoken Canadian artists when it comes to critiquing Canadian art and artists, was born in Brampton in 1925 and educated at Brampton High School. After his secondary school education, Bloore earned Master's degrees in art and archeology.

His view of art can be seen in a letter to *Canadian Art* in 1951, in which he lambasted Canadian artists for not progressing and for not aiming above the adequate. Bloore, who describes himself as a "simple painter," is constantly trying to change and improve his work, and his accomplishments attest to his respect for art. His studio is located not far from his home in Brampton.

William Ronald and John Meredith are both artists who, although not born in Brampton, lived many of their formative years here.

Born in 1926, Ronald moved to Brampton in 1940. He lived here for a time before going to New York. After he returned to Canada, he was the recipient of many awards, including the Guggenheim award for Canadian art in 1956.

Meredith was born in Fergus in 1933 and moved to Brampton at the age of six. After graduating from Brampton High School, he attended the Ontario College of Art. It has been said of Meredith's work that his drawings look painted and his paintings look drawn, which demonstrates the strong connection between the two forms. His works are on display in New York and throughout Canada.

For many, the fact that Brampton possesses a rich and diverse artistic heritage is something hidden, but it is there and is still expressed today.

—Ken Moore

An etching by Brampton artist Caroline Armington. Courtesy, Peel Art Gallery/Peel Heritage Complex

This is the Old Timers Hockey Team of 1966. From left to right: Murray Henderson, Brian Cullen, Sid Smith, Murray Eseard, Barry Gullen, and Bob Robertson. Courtesy, Guardian Collection, Region of Peel Archives

High School Wasps in 1882. Rugby eventually surpassed it in popularity, although in Brampton today, as in the rest of Canada, soccer is experiencing a remarkable renaissance.

Baseball arrived in the mid-1800s. It was introduced to Brampton by an American, George Kidd, who worked at the Haggert foundry. John Haggert, himself a keen supporter of sports, is believed to have brought lawn bowling to Brampton. Bowling, a popular sport in Haggert's native Scotland, was first played on the lawn at Haggertlea. The gardens were terraced, but the flat, lower part was perfect for this genteel sport. Brampton Bowling Club was founded in 1895.

Another favorite Scottish sport was curling. The Brampton Curling Club dates back to the 1870s. Early ice rinks were on the Etobicoke River or somebody's frozen backyard. A more permanent, covered curling rink was built by Fred Burrows in 1875. The members of the Brampton Curling Club became skilled players (as they are today), and in 1883 they won the Ontario Championship. This was no easy feat as they were up against stiff competition. *The Conservator* covered the event: "The Brampton curlers are champions of Ontario. Yesterday at Toronto they won the Governor-General's Grand Silver Tankard, valued at $600 beating the clubs pitted against them, the Barries and the Ancasters, these clubs having beaten all of the other of the eight districts in competition. . . ." Despite the primitive locales, many of these early teams did remarkably well. In the 1920s and 1930s, they went from strength to strength, and in fact, many teams are still in existence. Those early sportsmen would be amazed at the plethora of facilities that exist in Brampton today—heated arenas, squash clubs, tennis

The Brampton Excelsiors began as a high school team as early as 1871. These are some of the members of the Brampton Excelsiors of 1890 (left), and the Excelsiors of 1915 (below), who were coached by J. Carmichael. Courtesy, Region of Peel Archives

BRAMPTON FLYING CLUB

About forty years ago a handful of Brampton men got together and dreamed of flying, and even though most of them had never even been near an aircraft before, they formed a flying club that is still active in Brampton today.

In 1944 World War I veteran James "Mac" McCleave called some friends and asked if any would be interested in forming a flying club. The aviation seed had been planted.

The friends met in the old Registry Office (now the Peel Heritage Complex), and for many of those present, just the thought of forming a club of any kind was incentive enough. On this social note, the Brampton Flying Club was born—all that was needed was a plane to fly.

That problem was solved when club members learned that a war-surplus Tiger Moth was available for $250. Club member Bill Farr, the only one with a pilot's license, went to Oshawa and flew the Tiger Moth back to Brampton.

One of the first landing strips used by the club was a field beside McCleave's garage at the northwest corner of McMurchy and Queen Street West. In order to raise money for the club, flights around Brampton were given for two dollars a trip. Elmore Archdekin, one of the club's founding members, recalls a complaint being made to the council by someone who was upset with the noise the

flights made on Sundays.

The club was not able to obtain a license because the Department of Transport required that the club have two runways, one running east to west and another running north to south. The Dixie Cup site only had room for the former. So members moved their club to a farm owned by Rankin Kellam on Highway 7, where the Kodak plant now stands. However, there still was not enough room for both runways. Flying out of that site was made more interesting by the fact that there were hydro wires not far from the end of the landing strip that had to be avoided.

Finally, a vacant area owned by the provincial Department of Agriculture was located where the present-day OPP Training School on MacLaughlin Road is located.

Farr met Minister of Agriculture Tom Kennedy to seek the department's permission to use the land. Fortunately, remembered Farr, Kennedy had been up in a helicopter the day before, and thought flying was the greatest thing. He made arrangements for the club to lease the field for one dollar a year.

What followed was the difficult task of clearing a field with over 200 trees on it. Members worked long hours with equipment borrowed from club-member Charlie Armstrong's construction company, which had built runways during the war. Eventually, the toils of the club members paid off and the field was cleared. The club in 1946 obtained its charter as well.

The club operated well through the years until 1968, when they were notified that their lease would not be renewed. Their search for another site was made that much more difficult because of Brampton's proximity to the international airport. Eventually, a site was selected in Caledon, and the official opening of the new club occurred in 1970 with the first of the club's air shows.

Brampton Flying Club is now considered to be a world-class flying school, with twenty-one instructors on staff with the most up-to-date equipment at their service. The enrollment has increased, and the club has become much more professional, but the adventurous spirit of those early days of flying still survives.

—Ken Moore

Brampton's flying club was established in 1944, and by 1970 it began to stage exciting air shows that exhibited vintage biplanes and feats of flying by club members. Courtesy, Guardian *Collection, Region of Peel Archives*

courts, golf courses, soccer pitches, basketball courts, and baseball diamonds.

Brampton's favorite sport by far was lacrosse, Canada's national game. The sport originated with the Indians, and was named by French explorers because of the unusually shaped stick with which the ball is thrown. Whether or not it was played by the Mississauga Indians who once roamed Chinguacousy is a matter for conjecture, but the game as it is known today originated in the late-nineteenth century.

Lacrosse was introduced to Brampton by George Lee, a grammar school teacher, and was first played here in 1871. He gave the team the somewhat lofty title of the Brampton Excelsiors, but it proved rather appropriate

SPIRIT, MIND, AND VOICE

Right: A Brampton fireman in full uniform circa 1884. The volunteer force was headed by James Golding, the first fire chief. Photo by J.W. Cole. Courtesy, Region of Peel Archives

Facing page: Brampton's first official volunteer fire brigade was founded in 1883. These are delegates of the 10th Annual Convention of Provincial Volunteer Firemen, held in August 1909. The men in the uniform with "B" on their hats are the Brampton volunteers (see men at left). Courtesy, Region of Peel Archives

because the team went on to scale the heights of sporting greatness. The name Excelsiors evidently became a Brampton favorite. It was used for the rugby team, the ice hockey team, and even Brampton's first volunteer fire brigade.

The lacrosse team attracted a loyal following, and although the Excelsiors didn't always win, their playing was nearly always memorable. In 1876, a contemporary report relates how the Excelsiors, now a community team, played a spectacular twilight game against Brantford, using flaming, oil-soaked balls.

Throughout the late 1800s, many prominent Bramptonians, such as

99

Brampton's first fire engine took part in Brampton's annual Flower Festival, which in 1973 had a Centennial theme. Courtesy, Guardian Collection, Region of Peel Archives

Samuel Charters, Roswell Blain, T.W. Duggan, Grenville Davis, and W.F. Milner, participated in the game. In 1893 and 1894, the Excelsior Intermediates won the Ontario Championship. In 1914 the senior team made a succesfull bid for the covetted Mann cup in Vancouver. In 1930, 1942, 1972 and 1980 they brought the trophy home again.

The covered curling rink on George Street, so lovingly erected by enthusiasts in 1875, burnt to the ground in the late 1890s. Fires broke out fairly frequently in Brampton's early days, and in an era when there were no professional firemen, there wasn't much anyone could do about putting them out. Fires, along with primitive roads, harvest failures, water shortages, lack of education, and bad health, were just one of many obstacles that settlers learned to live with.

When a fire broke out, somebody ran to ring the town bell, then located at John and Chapel streets. Volunteer firefighters rushed to the bell to learn where the fire was—using up valuable minutes. By the time the volunteers got their inadequate resources to the fire, it was often beyond control.

The first "hook and ladder" company was formed by Norm McConnell in 1873. Hook and ladder companies got their name because of the way they fought fires. A large hook was attached to a rope, then thrown over the roof of a burning building. The building was hauled down, thus creating a firebreak that prevented flames from spreading. It was a primitive method that was not very effective.

Fire was not such a problem in rural areas. A family might lose their home, but because the homes were widely spaced, the fire was less likely to spread to other buildings. The availability of raw materials and helpful neighbours usually ensured that a house was quickly rebuilt. But in towns where buildings were close together, the consequences could be drastic. Brampton almost lost its main street in 1886.

The main problem in the early days was inadequate access to water —vital to the operation of a fire service. When Brampton was incorporated as a town, the communal water was drawn from shallow wells less than 50 feet deep. Their capacity was limited, and the only way to get the water that was there was with "old oaken buckets."

On March 17, 1878, an act was passed by the provincial government authorizing the "appointment or election" of three water commissioners to administer all matters relating to the supply of water. The 40-acre Snell's Lake (Heart Lake) was an obvious source, but a pipeline had to be laid to bring the water into town. After numerous arguments, M.M. Elliott (who lost a mayoralty race over the issue) persuaded property owners to authorize a line into Snell's Lake. The system worked well and Brampton's manufacturers were delighted because the increased water supply now furnished water power for their machinery.

Five years later Brampton's first official volunteer fire brigade was founded—The Excelsior Hose Company #1. James Golding was appointed chief. The new fire service and the new water system were soon put to the test when the Iron Block on Main Street caught fire, but both worked very well. The fire was contained within a small area and although the Iron Block was lost, the rest of Main Street was saved.

When horse-drawn engines replaced those pulled by hand, the fire brigade became even more efficient. Men could get to a fire more quickly than before, and bigger engines could be built with greater water-pumping capacities. Brampton's alarm system was improved in 1914 when fourteen street alarm boxes were installed. James Harmsworth (of the paint and wallpaper store) was fire chief at that time.

In those days, it was customary for the fire department to offer bonuses of one dollar to the fastest "reel men." And children, who usually found fires fun rather than frightening, often tried to get in on the act. Gordon Beatty, writing in 1973, reminisced:

We had a volunteer fire brigade with hand drawn fire reels stationed in each of the four wards (of Brampton). Those who reached a fire first with a reel were paid a dollar a piece. This created quite a bit of competition. Out of school hours many of us kids were on hand to help pull or push the reels, but our efforts were seldom appreciated. At any rate, we did not get paid. . . .

Brampton's fire service has come a long way. Today it is a high-tech operation. It has a modern communications system, efficient equipment, speedy vehicles, highly trained personnel, and last, but certainly not least, municipal funding—a far cry from those early days when many a Brampton home was lost to fire and smoke.

These are members of Brampton's famous "Flower Family," the Dales, photographed at the Algie House on Highway 10 circa 1898. William Algie, Harry Dale's son-in-law, is third from the right, at top. Billy Little, far right, worked at the Dale Estates for fifty-two years. Harry Dale (with hat) is standing to the right of the pillar at left. The house was demolished in the late 1970s. Courtesy, Region of Peel Archives

CHAPTER FIVE
THE SHAPE OF GROWTH

As Brampton entered the twentieth century, it became less isolated. An up-to-date transportation and communication network linked it with the rest of the country and Brampton was no longer just the county town, revolving within its own sphere of influence. What went on in the rest of Canada involved Brampton as never before—politically, socially, and economically.

The country was going through many changes. The years following Confederation had given Canada a sense of nationhood, and she was pulling herself out of the colonial era. When the Boer War broke out in 1899, there were numerous discussions in Parliament about whether Canada should, or should not, support the Mother Country. Many Canadians felt that Canada was now autonomous and therefore should not concern itself in an imperialist war. There were emotional demands from English Canadians to rally around the flag. There were equally strong demands from French Canadians who did not want to be sent to die in an English colonial war—a theme that was to recur in World War I when conscription was enforced.

For Bramptonians, however, these arguments were hardly an issue. The town, after all, had been founded by the British. To them England was still the Motherland—emotionally as well as legally (as a member of the Empire, Canada was automatically involved in the war)—and when Britain declared war on Germany in August 1914, Bramptonians were quick to rally around the flag.

The economic highs and lows brought on by World War I—the Roaring Twenties and the subsequent Depression—affected Brampton as it did the rest of the country, but once again, Brampton's diverse economy cushioned it from the worst of the swings. From the turn of the century to the end of World War I, there was very little increase in Brampton's population. In 1895 the population was 3,070. By the end of the war it stood at around 4,000.

But World War II—a conflict that required more technology and equipment and hence more factories and people to run them—gave a boost to Brampton's industry and caused the population to rise. By 1948, 6,783 people were living in Brampton, and more than 32 percent were involved in manufacturing.

Prior to World War I, many Canadian firms had

Brampton's Queen's Hotel offered accommodations to travellers in the early 1900s. Automobiles were beginning to appear in town, signalling Brampton's greater link with the rest of the nation. Courtesy, Region of Peel Archives

already started to think of decentralization (by no means a modern idea). Several companies moved their operations to Brampton. Firms involved in a wide variety of enterprises, such as Gummed Papers Ltd., the Brampton Pressed Brick Company, The Williams Shoe Company, Charters Publishing Company, and many more, set up shop. Copeland-Chatterson, the first Canadian company to make loose-leaf office binders, opened for business in 1905.

One of the main reasons such companies were attracted to Brampton was the town's willing and available workforce. The demise of Haggert Bros. had left a gap in the market because, while other companies had come to fill its place, none were as large as the foundry.

Researcher Michael Proudlock has noted: "Although many of these industries did not require a large percentage of skilled labour, and those skills which were required could be readily taught, the fact that Brampton possesses a sizeable labour pool (200-250) of men accustomed to working with machinery in the Haggert factory, was an added attraction."

Labor was also very cheap. The workforce was not unionized, and teenage boys, some as young as fourteen, were given employment at rock-bottom wages even for that time. Those that had good jobs were lucky to get five cents an hour. The older ones got more, but rarely more than ten cents an hour. Those

THE SHAPE OF GROWTH

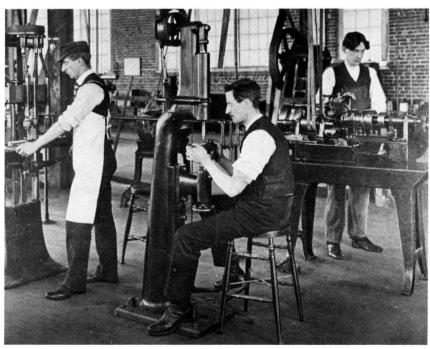

Prior to World War I, many Canadian firms had already begun to move toward decentralization. Several companies moved to Brampton. Copeland-Chatterson, the first Canadian company to make loose-leaf office binders (interior view, right), opened for business in 1905. Courtesy, Region of Peel Archives

105

This photograph of Boyle's Drug Store interior was made in 1942. Courtesy, Region of Peel Archives

who did not work in the factories picked up odd jobs here and there because even children were expected to help support their families.

Gordon Beatty, writing in 1974, remembered that as a boy he had worked as a cowherd:

I had to drive Dave Smithers' cows half a mile down Main Street and back, for milking morning and evening. For this chore I was paid the magnificent sum of two cents per day.

We youngsters seldom got more than a few coppers but of course the coppers of that day—larger than present day quarters—had considerable purchasing power. One copper would usually buy three or four Quaker molasses kisses.

Beatty also recalled that the presence of so many factories caused considerable pollution to the environment: "Our air was not much polluted by the exhausts of motor vehicles since there were very few of them but smoke belched from the chimneys and considerable sewage drained into our creeks."

THE SHAPE OF GROWTH

Above: Robert Dale, a member of the Dale "Flower Family." Courtesy, Cole Collection, Region of Peel Archives

Top, right: Harry Dale, circa 1880. Harry Dale's father, Edward, grew vegetables for door-to-door sales, and Harry joined the business in the 1870s. Under his management the company grew to become a major flower producer. From A History of Peel County

Bottom, right: T.W. Duggan became manager of the Dale Estates following the death of Harry Dale in 1900. Duggan also became Brampton mayor in 1912. From A History of Peel County

Dale Estates Nurseries was Brampton's largest employer in the early 1900s. The flowers that made the Dales, and Brampton, nationally known were grown in these greenhouses. Courtesy, Region of Peel Archives

Facing page: This is the Stork family circa 1890 at the house that is now the Ward Funeral Home. At left is Edwin Stork (1853-1921), who operated the pharmacy next to the Royal Bank on the Four Corners. Next to him is William Stork, a veterinarian. Louis Stork (far right) operated the Robinson and Stork General Store. Courtesy, Region of Peel Archives

Nevertheless, dirty water didn't stop youngsters from swimming in Fletcher's Creek. The Grand Trunk railway tracks ran alongside it and the company apparently often got complaints from passengers because boys were swimming there in the nude—not an acceptable practice in a prudish Victorian society.

Brampton's largest employer at this time was the Dale Estates nurseries, which, to some extent, took over where Haggert Bros. left off. The company began in a modest way in 1860. Edward Dale, a newcomer from Dorking, England, started growing vegetables, which he sold from door to door. In those days he had only one small greenhouse. His son, Harry, joined the business in the 1870s. Harry's particular enthusiasm was roses, and under his management the company grew from strength to strength. Harry Dale died in 1900 and the company, now a major concern, was taken over by T.W. Duggan (later mayor of Brampton) and William Algie, Harry Dale's son-in-law. By 1905, Dale Estates Limited had 150 employees.

Dale Estates was one of the first nurseries to popularize orchids. By mass-producing these exotic blooms they were able to bring their price down, so much so that the man on the street could now afford to buy them. In 1934 Harry Algie developed a method of punching holes in rose leaves. The holes could be punched to form letters for advertising slogans and this gimmick quickly put Brampton's flower industry on the map.

The "autographed" roses won prizes at flower shows in Detroit and New York and they were shipped to every corner of the globe. (In 1953 the company sent one hundred roses to Queen Elizabeth II for her birthday.) Dale Estates also experimented with new, exotic varieties of roses like Canadian Queen,

THE SHAPE OF GROWTH

Lady Canada, Lady Willingdon, Rosedale, and Sunbeam. As well they grew lilies, chrysanthemums, daffodils, and tulips. Brampton became known as Flower Town—a nickname that has stuck. The City of Brampton still uses a logo that incorporates a rose, with the slogan "Brampton is Blooming."

Satellite industries, which depended on Dale's for their livelihood, grew up around the company. Some of the people who had learned their trade at Dale's went on to found nurseries of their own, most notably the Calvert family (later Dale's became Calvert-Dale). At the flower industry's peak in the 1950s, Brampton had more than a dozen wholesale florists, including the Fendley and the Jennings families, who had worked in the floral business through several generations. In 1953 alone, Brampton's flower growers produced over 14 million blooms.

But at the start of World War I, most businesses were much smaller than Dale's. There were numerous family-run shops. Brampton's oldest stores —Blain's Hardware and Harmsworth Paint and Wallpaper—were going concerns. Blain's was operated by Richard Blain, the well-known politician. After concentrating on the store for a couple of years, he eventually went into politics full time. During his time in office, he established, among other things, postal delivery in Peel County. He introduced mail boxes and was instrumental in the construction of Brampton's Armoury.

When Richard Blain became a politician in 1912, his son Roswell took

over the hardware business. Roswell was an avid sportsman. As well as running the store, he played with the Brampton Excelsiors. He was a member of the lacrosse team that in 1914 made a bid for the Mann Cup. In 1953 Roswell sold the store to Lloyd Wagg. Today Blain's Hardware is run by Lloyd's son, Lewis Wagg.

Harmsworth's Paint and Wallpaper has been operated by the Harmsworth family since before the turn of the century. It was bought by James Harmsworth, a butcher, who passed it on to his son, also called James (who was also Brampton's fire chief for twenty-five years). Elmer Harmsworth, grandson of the founder, is still alive, and he recalls that during his father's day, "Harmsworth's probably painted every church and school in the town." (At that time the store also hired out painters). The store is currently run by fourth- and fifth-generation Harmsworths—Jim Harmsworth and his son David.

Harmsworth's Paint and Wallpaper, seen here circa 1920, has been operated by the Harmsworth Family since the turn of the century. James Harmsworth (left) is standing with a Mr. Street, a commercial traveller. Courtesy, Region of Peel archives

The Hewetson Shoe Company was another family-run firm. In size it ranked second only to Dale's. Founded in 1908 at the corner of Terauley and Bay streets, it was one of those Toronto companies that moved to Brampton for economic reasons. John W. Hewetson and his son A. Russell Hewetson were doing well purchasing offcuts of leather from men's shoe factories, and then turning the offcuts into shoes for children. The thrifty founders decided they could do better if they moved to Brampton.

The Hewetsons were so keen to get going that they moved the machinery and equipment to Brampton before the new factory was finished. With the help of carpenters, millwrights, and machinists, everything was installed in record time. The factory opened for business in January 1914, at 57 Mill Street North. By the end of that year, the Hewetsons had fifty employees who turned out 240 pairs of shoes per day.

The company suffered a temporary setback in September 1916. Fire gutted

THE SHAPE OF GROWTH

This is the Camp Kitchen Carnival Parade along Main Street North in 1915, just after the start of World War I. Brampton, with its British background, was strongly patriotic towards the Mother Country. Courtesy, Region of Peel Archives

the top half of the building, and although Brampton by now had a volunteer fire service, considerable damage was done before the men could get there. The fire had broken out during the night and because Hewetson's didn't have a watchman the fire wasn't discovered until it was well under way. The cutting room was completely gutted. (Brampton was to experience another major fire the following year when the high school burned down.)

The cutting room was soon rebuilt, but something equally serious hit the factory shortly thereafter—a chronic shortage of manpower. Bramptonians, like other Canadians, were leaving for the war in Europe in ever increasing numbers, and for awhile Hewetson's had to move part of production to New York State.

World War I was to have a profound effect on Canada, not just in terms of lives lost, but also in terms of the economy. The backwash of this war was followed closely by the waves of change wrought by the Depression and then World War II, together reverberating through the twentieth century until well

This regimental piper was a member of the 20th Lorne Rifles, outfitted in 1881 dress. Peel County sent soldiers to Europe during World War I in the body of the Lorne Scots. The regiment originated in the 1790s when the first militia was raised in what was then Upper Canada. Courtesy, Region of Peel Archives

into the 1950s.

Militarily, Canada contributed heavily. The Canadian Army eventually required more than 600,000 men. Some 8,000 men were enlisted in the Royal Navy. *(The Conservator* ran many advertisements for volunteers.) Twenty-four thousand went into the British Air Services. The effect on the economy was startling. Demands for farm produce and raw materials were almost unlimited, and during the war, exports of grain and flour doubled in value. Raw materials such as wood, pulp, meat, livestock, and metals pushed up export figures to unheard-of levels.

World War I was announced by *The Conservator* on August 4, 1914, in ominous bold type:

BRITISH EMPIRE AND ALLIES ARE AT WAR WITH GERMANY

The British Empire, of course, included Canada. In Brampton, feelings about German aggression ran high. A few days before war had been officially

THE SHAPE OF GROWTH

In 1936 the combined regiments of Peel, Halton, and Dufferin became the Lorne Scots, headquartered in Brampton. This ceremony was the Laying Up of Colors on April 30, 1939. Courtesy, Region of Peel Archives

proclaimed, Richard Blain, then a member of Parliament for Peel County, made known his views in the *Montreal Star*: "Should Great Britain be involved in the threatened European war, Canada should offer the thirty-five million dollars voted by the House of Commons for three dreadnoughts, and a first volunteer contingent of one hundred thousand men. Peel County will freely offer her quota."

Peel County did indeed "offer her quota" in the body of the 36th Peel Regiment (the Lorne Scots). The regiment originated in the 1790s when the first militia was raised in what was then Upper Canada. In the 1860s, after the Fenian Raids, the Canadian government decided to update and reorganize the country's army. Under a general order dated September 14, 1866, the 36th Peel Battalion (later the 36th Peel Regiment) was formed. On September 18, the 20th Halton Battalion of Infantry was formed.

The "Scottish connection" came about when, after reviewing the Halton Rifles on September 27, 1879, the Marquis of Lorne (Governor General at the time) bestowed on the soldiers the right to wear kilts, trews, and glengarries—black woollen caps with distinctive red and white checkered

113

BRAMPTON

Brampton's knitting mill (seen here circa 1930) is now a restaurant and office complex. It started life as a lumberyard in the late 1800s, and at the turn of the century it became a knitting mill. Courtesy, Region of Peel Archives

bands. In 1936 the combined regiments of Peel, Halton, and Dufferin became the Lorne Scots (Peel, Dufferin, and Halton Regiment), headquartered in Brampton.

When World War I broke out, Canada sent an expeditionary force to Europe, and as the war progressed, draft after draft was called from Peel County. These men were engaged in some of the most important battles of the war—the second battle at Ypres, the counterattack at St. Julien, and the appalling Battle of the Somme. The sons of many prominent Brampton families went off to fight. Roswell Blain, son of Richard, served overseas with the 126th Niagara Battalion. He fought alongside another Bramptonian, David Bovaird, a member of a local horsebreeding family. Two sons of Samuel Charters joined in the fray. Harry and Reg Charters saw combat in the infantry. (Grandson Sam Charters enlisted with the Lorne Scots in World War II.) Earl and Russell Harmsworth, James' sons, left in 1916. Sadly, Earl was killed in France, only months before the war ended.

Hardly a day went by without somebody going to the front. As the war ground on, news trickled back, not all of it pleasant. *The Conservator* read like a roll call of disasters:

"Missing since October last, Private Letty is now killed in action, according to word received from Ottawa. He had been in the trenches a short time before taking part in the big battle on the Somme in which he lost his life. He was 21 years of age and an only son. . . ."

A Private T.A. McClintock wrote ruefully: "I am afraid the old town of Brampton's death roll will be pretty high before the war is over but nobody can say that the town did not do its share. . . ."

The loyalty and dedication shown by these young men was truly astounding. One *Conservator* article talked about a silver watch that had been brought back to Brampton and given to a soldier home on leave by one Sidney Graves, who had taken it as a souvenir from a dead German soldier. *The Conservator*, July 5, 1917: "It is probably true that none of the men going from Brampton have had greater experience in the many battles than has Mr. Graves. He has been wounded but is again in the trenches doing his part."

Some of the letters sent by soldiers were downright bloodthirsty, perhaps not surprising, given the extreme patriotism of the times. Private David Bovaird, after being wounded for a second time, wrote to his family:

I am in the best of health outside of a nice little lucky wound in the left shoulder that I received with pleasure in the advance at Lens and Loos. We sure had some time and I gave them the benefit of my rifle and ammunition.

It was my last chance to avenge Alex McClelland's death and I certainly got some of them before they got me. We were fighting out in the open with machine guns and rifles when we advanced on them. They opened rapid fire on us, but we didn't worry about that. All I was thinking about was filling them full of lead. . .

But such bravado often hid fear, loneliness and acute homesickness. In a letter written by R.K. Brown on July 4, 1917, he asked wistfully:

THE WOMEN'S INSTITUTE

At the turn of the century, Brampton's farmers exchanged ideas and learned from others' experiences through the Farmer's Institute, a forum they had set up to expand their pool of knowledge. But the farmers' wives and other area women had no such organization until Robert McCulloch decided in 1901 that women were entitled to an opportunity to take part in community affairs and widen their knowledge at the same time.

The idea had come about a half-decade earlier in the tiny village of Stoney Creek. Women in that area had attempted to form some type of women's club, where the wives of farmers could exchange ideas about the home and family to overcome the occasionally tragic consequences that resulted from leading lives in near isolation from each other.

One of their early meetings included a talk by a Hamilton mother who had lost a young son through the feeding of impure milk. This account brought home to the women the effects of the isolation that was felt by them all. It was at that moment that the initiative was taken and the first Women's Institute was formed. The idea would spread far and wide.

The first Institute in the Peel area was in Brampton. McCulloch, who was then secretary of the Farmers' Institute, made sure that while the men were watching demonstrations on the latest farming techniques, their wives would be able to attend the great curiosity that was Ida Hunter's travelling kitchen. Hunter was from the Department of Agriculture, and her demonstrations were so impressive that she was named the secretary for the Women's Institute of Peel.

The Peel chapter held their first meetings in the Concert Hall, and after the first year, the group elected officers—Mrs. George McClure became president. With the formalities out of the way, work commenced on developing the chapter's mandate: to bring together town and country women; to develop confidence in timid women; to broaden women's thinking through the exchange of ideas; and to add interest to women's everyday work.

As time went by, more institutes sprang up in Brampton and its surrounding countryside. As the institutes grew, so did an interest in improving the community, and issues of social concern moved to the forefront.

As early as 1909, the need for a hospital had piqued the interest of the women. By 1910 a charter was granted to the Institute for the Peel Women's Institute Hospital. In 1922 the charter and funds were passed to the Board of Governors for Peel Memorial Hospital.

Today, women are finding outlets in the workforce in ever-growing numbers, and the seven branches of the Women's Institute in the Peel South District, as are many of the branches now located in urban areas, have found membership waning. The Institutes in Brampton may soon become a thing of the past. This, however, has not weakened the determination of the 250 to 300 members in the Peel South branches. Nursing home care and health care costs are among the concerns voiced by the Institutes. In addition, the Tweedsmuir histories kept by each institute have become invaluable sources of information for those interested in local history. While there may come a day when the Women's Institute will no longer function, its imprint on the community's maturation will remain a lasting legacy.
—Ken Moore

Prior to 1901, women operated households without benefit of a forum for exchange of health and child-rearing information, with sometimes tragic results. The Women's Institutes provided such a forum, and today its concerns include elderly care, health costs, and local history.

THE SHAPE OF GROWTH

How is sport in the old town this summer. The boys out here have gone in for quite a bit of sport. The running high jump was won at 5 ft. 1 in. The winner gave an exhibition at a greater height. Then followed the obstacle race, relay, tug-of-war and bomb-throwing.

We all made our way back feeling that we had spent a memorable day in the old Canadian way, but a long way from The Land of the Maple. . . .

Meanwhile, in Brampton, citizens were doing all they could to help the war effort. The British government needed funds to support their armed forces; in 1917 Ottawa levied its first income tax to raise "Victory Loans." *The Conservator* was full of advertisements for food, artillery horses, men, and money. Citizens were exhorted to come to England's aid. "For Canada it is both a filial and patriotic duty to supply Great Britain's war needs," thundered one ad. "No task is too great—no obstacle insurmountable. WE are CANADA and SHE is GREAT BRITAIN - both members of the same great Empire, kin of our kin, our Motherland," bellowed another.

The Department of Agriculture urged farmers to grow extra crops as a "high, patriotic and unselfish service" to "their brothers overseas." Bramptonians were told to register with J.W. Stark, the District Representative for the Department of Agriculture. By the time the war ended, Canadians had raised two billion dollars—a phenomenal amount of money for that era.

Few people were exempt from the drive (or desire) to help Mother England. Farmers grew bumper crops. Factories worked at full steam (the Brampton Knitting Mills churned out socks for the soldiers) and charitable and church groups sprang into action to help fuel the cause.

Numerous "Patriotic Outings"—picnics, dances, garden parties and the like—were organized to raise funds. Self-help groups (though they certainly weren't called that then) got together to learn about canning and preserving food and they sent what they could overseas. Evidently it was much appreciated. One soldier wrote that "the young ladies of Brampton do not forget us boys," although he did lament that while they sent a great many "useful articles" they omitted to send whisky.

Among the most active groups were the Women's Institutes, who worked tirelessly on fund-raising schemes. Before war broke out, they had started raising money for a hospital (Brampton still didn't have one). With the help of clergy, members of parliament, doctors, lawyers, and businessmen, the Institutes had managed to collect $2,263. The war put the project on hold but it also spurred the fund-raising effort once the conflict was over. It was decided, in fact, that a hospital would be a fitting war memorial to the men who had fought in Europe. In making their decision, a committee made up of the Brampton Branch of the Great War Veterans Association, the Peel Women's Institute, the Old Countrymen's Club, and the Brampton Cheerio Club passed the following motion: "whereas it is deemed advisable that this memorial take the shape and form of something permanently useful to the town and surrounding county as well as commensurate with the sacrifice made. . . ."

Land was found and plans to build got underway. The William Elliott estate was deemed most suitable and after raising $9,000 (some of which was

BRAMPTON

Peel Memorial Hospital (seen here in 1973), opened in 1925 and underwent almost continual expansion due to overcrowding in the decades from the 1930s through the 1970s. In 1986 hospital administrators were once again seeking funds for renovation and expansion. Courtesy, Guardian Collection, Region of Peel Archives

left over from the Peace Day Fund) the sale went through. The Peel Memorial Hospital opened in 1925. Free medical care was given to former servicemen and women and to wives of men killed in action. The building had six private rooms, two public wards (with half-a-dozen beds) and a nursery big enough to accommodate three babies.

Right from the start the hospital was a great success; one hundred fifty-six patients were admitted in its first year of operation. Because it was the most advanced facility in the area at the time, the hospital fell victim to what has become a recurring problem—overcrowding. Additions were made in the 1930s, 1940s, 1950s, 1960s, and 1970s, and in 1986 hospital administrators were once again seeking funds for renovation and expansion.

Although Peel Memorial Hospital was not completed until after the war was over, other moves had been made to safeguard public health. Brampton's water supply at Snell's lake was proving inadequate for an increasingly industrialized town, and in any case, the water was not particularly clean.

In 1912 a new reservoir—Brampton's first—was constructed. Built at a cost of $22,500, it was capable of storing one million gallons of water. Six years later, Brampton became one of the first towns in Ontario to install a municipal sewage treatment plant, using the activated sludge treatment, a method that gave a high degree of purification. By 1920 chlorine was added

THE SHAPE OF GROWTH

Above: The advent of the automobile brought many changes to Brampton, including better roads. This is Main Street (looking south) as it looked in 1920. Courtesy, Brampton Public Library

Right: This is the Etobicoke Creek in 1907, at a spot opposite a Brampton park. Courtesy, Region of Peel Archives

to Brampton's water, a move that greatly reduced the risk of illness.

On November 11, 1918, *The Conservator* announced—in four-and-one-half-inch high letters—the triumphant news:

ARMISTICE IS SIGNED

Bramptonians were ecstatic. Factory whistles screamed. People honked their car horns and took to the streets, shouting and singing with joy. "Never before," said *The Conservator*, "did jubilation run so high in the county town." Little did Bramptonians know that a similar scene would unfold less than three decades later.

The war, which had given such a spurt to industry, also brought a boom in the following years. Urbanization and industrialization was making itself felt. People were moving to the cities to work for higher wages and consequently

had more money to spend than ever before. This period of prosperity had a number of intrinsic weaknesses that later led to an economic slump—the worst in modern times—but meantime, the country was in vigorous health.

By the mid-twenties Canada was enjoying a period of unabashed consumerism. Radio and motion pictures had just made their debut and they quickly became the most popular forms of entertainment. Auto touring became fashionable, particularly among the middle classes and upper classes who could afford it. A Brampton newspaper advertisement for Ford cars, although loaded with patronizing prose, conjures up that carefree era very well: "Thousands of wives and daughters run their own Ford cars," ran the heading. "They use

THE SHAPE OF GROWTH

Facing page: Motorized vehicles were slowly taking over in the 1930s, but horse-drawn delivery carts were still in use Courtesy, Region of Peel Archives

Above: Brampton's first telephone building opened on John Street in 1924. It heralded a change-over from the old "crank" system to an automatic system. The 1,000th telephone was installed the following year. Courtesy, Brampton Public Library

BRAMPTON

The interior of Brampton's telephone exchange in 1937; these operators are seated at a common battery switchboard. In the early 1930s, underground cables replaced above-ground poles, as the phone system underwent continued modernization. Courtesy, Brampton Public Library

THE SHAPE OF GROWTH

Following World War I, Brampton's economic base had shifted from agriculture to manufacturing, but increased mechanization allowed fewer farmers to produce more food crops than ever before. Mechanization began in the mid-1800s when horse-powered machinery replaced the slower oxen, and before the turn of the century, steam power replaced horsepower. Gasoline engines came into use in the early twentieth century. From A History of Peel County

The Bull Dairy Farm's new milk truck at Fan's Garage on Main Street in 1925. Mechanization and modernization enabled the Bulls to operate a large cattle and dairy business. In 1900 the Ontario farmer could feed himself and seven others. By 1966 the "others" had increased to over thirty. Courtesy, Region of Peel Archives

123

them for shopping, calling, attending the theatre, taking the children for a run in the country or to school." The ad went on to say that the Ford "was as easy to operate as a kitchen range," and that "no knowledge of mechanical details" was necessary. A "Touring Ford" sold for under $500.

Part of this new consumerism ran to telephones. Everybody wanted one. In 1924 a new telephone exchange building was erected on John Street. A "common battery" telephone system was installed, putting an end to the old hand-cranked phones. Five years later, Brampton had 1,302 subscribers —almost a third of the population.

Brampton's economic base had shifted to manufacturing but the surrounding rural areas certainly weren't idle. The number of farms in the townships of Chinguacousy and the Gore had certainly shrunk, but those that remained were steadily becoming more efficient. The farmers were able to produce more than ever before and the war had created an almost insatiable demand for meat and dairy products.

The County of Peel Agricultural Society furthered its policy of farmer education. Between 1910 and 1930, the Society ran rural youth programs—the forerunners of the 4-H Clubs. The programs were aimed at teaching youngsters how to raise animals and grow crops. Schoolchildren were issued plants and seeds and encouraged to raise calves, colts, and chickens.

It was during the "Roaring Twenties" that a famous Brampton cattle-breeding family—the Bulls—came to the fore. Flamboyant and wealthy, with more than their share of eccentrics, they lived in a free-wheeling, free-spending style. They hosted lavish parties atttended by people from around the world. Their sixteen-room mansion, Hawthorne Lodge, was surrounded by landscaped lawns and lily ponds, and was *the* social centre of town.

The Bulls were particularly renowned for their purebred Jersey cows. The first herd was established in Brampton by B.H. (Bartholomew Hill) Bull in 1874, and for many years he was the leading expert in his field. Two sons, Duncan and Bartley (and later his grandson John) took over the family business, and, with a third son, Perkins, helped to put Brampton on the map. Outspoken and colorful, Perkins promoted the Jersey herds to the many people he knew. He left farming to become a historian and writer.

The descendants of Bartholomew Hill were a distinguished lot. Jeffrey Bull was a major in the 75th Canadian Infantry. He was killed in action on August 8, 1918. The same year another Bull, Louis, was elected Brampton's mayor. He was followed by Bartholomew Harper Bull—who was Brampton's mayor in 1944 and again in 1952.

By 1925, the Bulls owned a mind-boggling 400 hectares of land and their Jersey herd had expanded to 1,000 animals, believed to be the largest Jersey herd in the world. The cattle were housed in sixteen barns run by fifty farmhands and herdsmen and their annual feed bill ran around $70,000. The Bulls held annual auctions, and using the slogan, "You can Win with Brampton Jerseys," they attracted thousands of spectators and buyers. They also hosted ploughing matches. The four-day match held in 1924 attracted 72,000 people.

The Bulls continued in business until well after World War II. In fact, the Bulls helped to replenish Britain's depleted Jersey herd. Even today few Jersey herds cannot trace their beginnings to sires of the Brampton strain.

THE SHAPE OF GROWTH

The halcyon days came to an abrupt end in October 1929 when the New York Stock Market crashed. The panic spread to Montreal and Toronto. On October 29 frantic stockholders tried to offload their shares at rock-bottom prices. In Toronto 331,000 shares changed hands; in Montreal, 525,000.

The stock market debacle signalled a widespread economic depression that was to last for almost a decade, and Canada was among the most vulnerable victims. With a semi-developed industrial system, the country was heavily dependent upon foreign trade and particularly dependent upon the export of grain and raw materials. During the "Hungry Thirties" Canada's GNP declined by forty-two percent.

Urban middle-class families weathered the economic tribulations better than most, but working-class families faced with unemployment and drastically reduced wages fared very badly. In an attempt to support their families, many men spent years wandering around the country, jumping on trains in pursuit of jobs, which in most cases didn't exist.

In the fall of 1930 alone, about 100 people left Brampton to find employment elsewhere. By 1932, more than a third of the workforce was out

The Conservator *building on Queen Street West. The* Conservator *was one of Brampton's major newspapers. It began publication in 1873, and was run by the Charters family for many years. In 1953 it was sold to Thomson Publications. Courtesy, Brampton Public Library*

PERKINS BULL

When it came to attaining a distinguished level of success, Brampton's Perkins Bull was one of those who was never satisfied. His colorful life has left both a valuable legacy and some interesting stories for Bramptonians today.

Eldest son of Bartholomew Hill ("B.H.") Bull, who built up one of the most prosperous Jersey farms in the world on the land that is now Peel Village, Perkins was definitely not in the same mould as the rest of the family.

Perkins Bull was born in Downsview in 1870 and moved to Brampton four years later with his family. On the death of his father in 1904, Bull inherited the operation of the farm. Owning a world-famous farm, however, was not to the liking of the young man, who had obtained his law degree from Osgoode Hall. It was during this time that a hint of things to come for Bull occurred when he was accused of forging his brother's name on a note for $12,500.

After leaving the control of the farm to his brothers, Bull decided to move to England to see what he could accomplish there. His outgoing manner and over six-foot stature made him a very memorable figure, and before long he had become a top-ranking international lawyer. His dream for a knighthood, however, was never realized.

Bull made quite a name for himself when he founded the Perkins Bull Convalescent Hospital. He was roundly praised for his generosity until an investigation revealed that he was receiving per capitas from each patient and collecting donations from wealthy Canadians.

Also about this time, Bull devised a money-making scheme whereby he sold English bog land to elderly Canadians back home, and Canadian tundra property to those in England. Prosecution was in the air, but the Bull family put up a $25,000 bond and the matter was dropped.

It was in Chicago in the 1930s that Bull started making headlines. He found himself working for a wealthy heiress who needed legal representation in a divorce proceeding against her husband. After the settlement, the husband sued Bull for $250,000, saying that Bull had had him under constant investigation, and the pressure had led to ill health. The suit was eventually settled out of court.

It was at about this time that Bull left the windy city under mysterious circumstances. The rumors that gangster Al Capone was after Bull were made a bit more believable when Bull's car crashed into a truck in Michigan, and instead of being taken to the nearest hospital, he was taken, seriously injured, to a hospital in Windsor 120 miles away. He stayed there only one night and then was transferred under guard to Toronto. No explanation was ever given for the moves.

During the years 1931 to 1938 Bull wrote a dozen books on a variety of topics, ranging from military history to reptiles to the history of Peel. Bull sent an army of researchers throughout the county and collected facts, folk tales, and gossip, which he wrote down but never titled or organized.

This quiet time in Bull's life was to end in 1938 when his friend from Chicago, the divorced heiress, was found dead in her home. Suspicion immediately fell on Bull when it was discovered that he was to be left a large sum from her estate. The investigation that followed showed that the woman had died of natural causes. However, her family battled Bull on his right to the money and eventually he settled for $250,000.

Throughout his life Bull became involved in just about every scheme he could. He had lumber interests in the Canadian West. Land purchased in Cuba eventually amounted to a 20,000-acre sugar plantation. He founded the Canadian Oil Company, which became another success. At one time he became interested in exploration and financed expeditions to the Antarctic.

As for his books, most are to be found in public institutions, to which he donated the majority of the small press runs. For his writing he was given an honorary membership in the Mark Twain Society of the United States.

Bull was constantly a man on the move. In an article about his many exploits, the *Toronto Telegram* in 1962 called him "Canada's most remarkable author to date." No matter what aspects of Bull's colorful life are analysed, there is no doubt that he was one of the most remarkable citizens that Brampton has ever produced.

—Ken Moore

Perkins Bull

C.V. Charters, son of Sam Charters who bought The Conservator *from A.F. Campbell in 1890. Courtesy, Brampton Public Library*

of work. The following year a fund-raising radio concert was held at the Capitol Theatre. It was broadcast over CFRB in Toronto and the proceeds were given to the Brampton Welfare Society, which distributed relief to the poor.

But for all that, Brampton was less affected than most Canadian communities. Just as had been the case some fifty years before, its balanced economy helped it to ride the economic storms. For many companies, it was business as usual, even in the depths of the Depression.

Dale Estates was still doing a roaring trade, as was the Hewetson Shoe factory. The shoe factory closed its doors for awhile but that was due to family illness. Russell Hewetson and his father, John Hewetson, died within two years of each other. Mrs. J.W. Hewetson took over the presidency and the company continued to flourish. (When she died in 1945, she was succeeded by A.G. Davis, father of William (Bill) Davis, former premier of Ontario.)

The publishing business also continued to thrive. In addition to printing

THE SHAPE OF GROWTH

The Peel Gazette *was Brampton's "Liberal" newspaper. The elections of the 1930s and 1940s were the subject of many lively editorials. Note the "low" telephone number; the* Peel Gazette *was among Brampton's early phone subscribers. Courtesy, Brampton Board of Trade*

The Conservator, the Charters Publishing Company put out the *West Toronto Weekly,* the *Weston Times and Guide,* the *New Toronto* (and Lakeshore) *Advertiser,* the *Port Credit Weekly* and the *Guelph Review.*

In 1932 the Charters family bought the *Banner and Times,* intending to merge it with *The Conservator.* This proved unnecessary because a new "Liberal" paper opened in Brampton—the *Peel Gazette.* Run by Dr. R.J. Hiscox, Ed Furness, J.O. Adams, and W.J. Foster, the new paper sparked off another period of political warfare in print, and the elections of the 1930s and 1940s made for some lively reading.

Meanwhile, in Europe, there were rumblings of another war. The imperial designs of the Germans were once more causing unease. The Depression of the 1920s had popularized Communism, mainly because it appealed to suppressed workers who were suffering from economic deprivation. But Communism was countered by extreme fascism, embodied by Benito Mussolini and later, Adolf Hitler.

In March 1935, the German government formally denounced the provisions of the Versailles Treaty. In direct violation of the agreement, Germany announced that it was reintroducing conscription and increasing the size of her armed forces. On September 1, 1939, the Wehrmacht ("defense force") invaded Poland. Two days later, Europe was again at war.

Once more Canada was expected to spring to Britain's defence. Loyally, *The Conservator* proclaimed:

> **THE EMPIRE IS AT WAR**
> **CANADA IS AT WAR**
> **PEEL IS AT WAR**

But despite these patriotic declarations, Canada did not rush to defend the Motherland, at least not initially. Even more removed from the colonial mentality, Canadians debated endlessly on whether or not they should get involved in what was essentially a European war. The subject sparked many debates in Parliament. Circumstances finally spurred Canada into action. As the war heated up and the real horror of Nazism unfolded, troops and funds were allocated to England.

Lt. Col. L. Keene, commanding officer of The Lorne Scots, was authorized to help assemble the First Canadian Division. Later the Lorne Scots were chosen to form defence and employment platoons for the entire Canadian army. They moved to Liverpool in January 1940. During the blitz of the city, they helped the authorities deal with civilian casualties. Major D.C. Heggie, medical officer, who took part in the rescue operations, was badly wounded. In recognition of his bravery, he was awarded the George Cross.

Throughout the war, more men served with the Lorne Scots than any other unit, and the regiment fought in every theatre that involved Canadian troops, except Hong Kong. A platoon of the Lornes served with the Queen's Own Rifles at the capture of Boulogne. Over half the men were wounded or killed. The 1st Division platoon landed on the beaches of Sicily on July 13, 1943, and portions of the 6th Brigade took part in the raid on Dieppe.

Once more *The Conservator* kept Bramptonians informed, bringing news from the battlefield and reporting on activities at home. In 1940 the paper won the Mason trophy for Canada's best all-round weekly newspaper, and the Williams trophy for the best editorial page. Its pages again carried news of numerous fund-raising campaigns organized by church groups, social clubs, and charities.

Huttonville Park, next door to Huttonville Dam, was a particularly popular spot for fund-raising events. The thirteen-acre site was leased from Tom Moorehead by partners Gord Vivian and James McCraken. They turned it into a recreation area for the public, complete with a large dance hall and a picnic area.

"There was excellent swimming," recalled Gord Vivian in 1980. "As well as local people, we got people from the city who came for a picnic and a singalong."

Singalongs—huge community get-togethers that attracted well over a

THE SHAPE OF GROWTH

Brampton's Glen Eagle Crescent crosses the Etobicoke River. The river was a favorite spot for fishing in days gone by, but it was notorious for flooding its banks. After the 1948 flood, a diversion channel was begun and was completed in 1953. This photo was made in 1940. Courtesy, Brampton Public Library

BRAMPTON

The runway at Malton (now Pearson International Airport) was laid by the Armstrong Brothers, a Brampton firm that is now a construction and horse-breeding conglomerate. Courtesy, Guardian Collection, Region of Peel Archives

thousand people—were held on Sunday nights. The crowds would watch the latest hit movies *(Gone with the Wind,* with Clark Gable and Vivien Leigh, had just been released) and during the summer big bands would play on the open stage under the stars. The main purpose of these social outings was, apart from having a good time, to raise money to send cigarettes (a luxury) to the soldiers overseas. After the movie was over, the names of local lads would be flashed on the screen and a hat was passed around. During the wartime years, the Huttonville singalong audiences sent one million cigarettes overseas. "It was the greatest thing," said Vivien. "I got 300 to 400 letters of appreciation from grateful soldiers."

Brampton launched a $15,000 fund-raising campaign on September 23, 1940. That was followed by several others, notably the Seventh Victory Loan Drive in 1944. The County of Peel raised an astonishing $4,900,000, and of that $670,000 came from Brampton, by far the biggest contributor. Brampton still exhibited a strong loyalty to Britain and all hands were on deck to help her survive and fight.

THE SHAPE OF GROWTH

American Motors (Canada) Ltd. is one of Brampton's major employers. They opened for business in 1956. This is the first Rambler off the production line. Courtesy, Brampton Public Library

By the time France was occupied, Canada had become a major arsenal and supply base for England. Over a million people worked directly in the war industry and Canada was producing thousands of vehicles—800,000 by war's end—in addition to ships, weapons, munitions, food, and aircraft.

Some of those aircraft were built at Malton (now Pearson International Airport), two-and-a-half miles from Brampton's boundary. The runway was laid by the Armstrong Brothers, a Brampton firm that is now a construction and horse-breeding conglomerate, and many Bramptonians commuted to work at the government-owned Victory Aircraft Limited, a practice that continued after the war when Victory was taken over by A.V. Roe Canada Limited.

World War II not only boosted industry, it prompted a flurry of activity in agriculture. Farm laborers were very much in demand, and farm "commandos" (adult men) were drafted to work in other parts of the province. Bramptonians rode to the Agricultural Office in Toronto in trucks loaned by large companies. They were paid between $2.50 and $4.00 per hour for their labor. To make up the shortfall, "Farmettes" and "Farm Cadets" (teenage girls

BRAMPTON

and boys) were encouraged by the Federal Government to work on the land during summer vacations.

Finally, on August 14, 1945, this war, too, drew to a close. Under a headline which read:

CROWDS GO WILD
CELEBRATING VICTORY AND PEACE

The Conservator reported:

A crowd unprecedented in Brampton history packed Gage Park yesterday afternoon to attend the Thanksgiving Peace Service. Prior to the service a parade formed at Roselea Park and paraded down Main Street to music of the Lorne Scots Pipe Band, Salvation Army Band and Brampton Brass Band. The parade was led by Mayor Harper Bull and Gordon Graydon M.P. followed by members of Brampton Council.

Three years after the war ended Brampton was hit by another disaster, this time caused by nature rather than man. On March 16, 1948, the Etobicoke Creek flooded its banks, inundating the town with about four feet of water.

Facing page: Floods were not new to Brampton in the 1800s; the Etobicoke had burst its banks many times since Brampton's founding, but nothing was done —until 1952—to prevent the floods from happening again. Courtesy, City of Brampton

Below: This is Main Street (looking north) after the flood of 1925, when the Etobicoke overflowed its banks and swamped the downtown district. Courtesy, Region of Peel Archives

THE SHAPE OF GROWTH

BRAMPTON
TIMES EXTRA!

Friday, August 28th, 1857.

IMMENSE FALL OF RAIN!

BRAMPTON FLOODED!

Last night such an immense fall of rain took place, that, early this morning, the River Etobicoke rushed down with fearful velocity, and so overspread its banks, that the greater portion of Brampton was flooded. Through the two railway bridges the water rushed into the principal streets, which were soon like rapid rivers. The water, in several places in the village, was above five feet deep. It went in at the windows of some houses. The damage done is considerable. The planks and sidewalks of some of the streets have been torn up, and small bridges in the neighborhood carried away. One house has been thrown on one side by the violence of the torrent. Business has been entirely suspended. The flood is now decreasing rapidly; so, by to-morrow, we expect Brampton will assume its usual appearance. It is acknowledged by all that this very unexpected immersion of our village is the worst yet experienced.

On March 16, 1948, the Etobicoke Creek flooded its banks, inundating Brampton homes and businesses. A diversion channel would be built by 1952, halting the destruction caused by the frequent flooding. This photograph was taken at 24 Main Street, looking east. Courtesy, Region of Peel Archives

THE SHAPE OF GROWTH

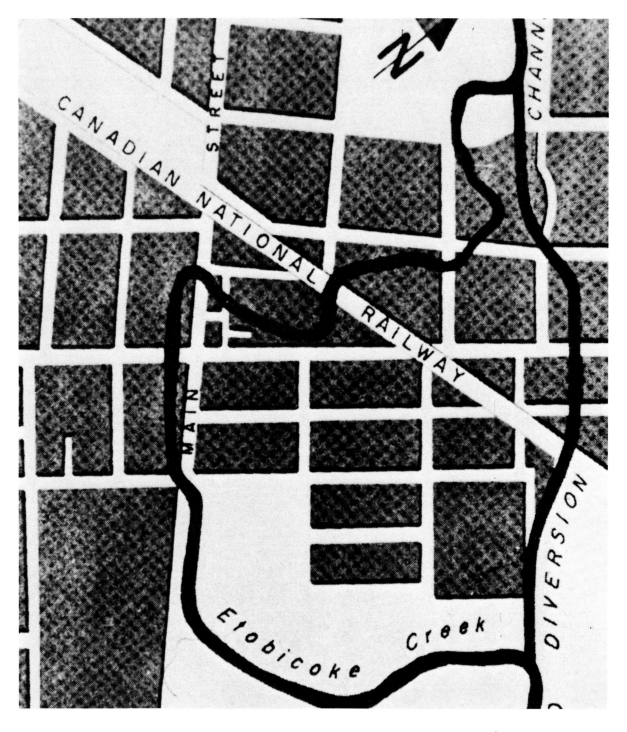

The flooding of the Etobicoke Creek in 1948 caused half-a-million dollars in damage, providing the impetus to construct a diversion channel. The channel ran in a straight line that quickly carried potential floodwaters away from downtown. Courtesy, City of Brampton

BRAMPTON

The flood of 1948 was particularly bad, as this view of Queen and George streets (top) and Main Street North (bottom) shows. When waters receded (facing page) much damage was revealed and the huge job of cleaning up and rebuilding began. Courtesy, Brampton Public Library

THE SHAPE OF GROWTH

Floods were not new to Brampton, however. The Etobicoke had burst its banks many times since Brampton's founding but nothing constructive had been done to prevent the floods from happening again.

The floods occurred because the Etobicoke swung through the centre of Brampton in an S-shaped path. The curved banks had the effect of slowing the flow, but when the spring ice melted, the volume of water increased and the creek took the path of least resistance—straight across the middle of town.

Given that this was a well-known problem, it was surprising that nothing had been done to prevent it. In 1854, after the entire downtown area was flooded, the village council merely commented on the "inconvenience of an overflow of water on Main Street." A torrential downpour on August 28, 1857, caused yet another flood. Again nothing was done. That flood was followed

THE SHAPE OF GROWTH

The coming of the railroads turned Brampton into a boom town. This is a steam train of the Canadian National Railway at "The Diamond" in Brampton in 1958. Courtesy, Brampton Public Library

by another in 1873.

Floods continued to plague the downtown area for the next fifty years. In 1912, Archie McKechnie, the ten-year old son of a local lawyer, was drowned. Six years later, James Crawford, a local funeral director, lost his life. C.R. Magee, a Brampton druggist, was a fatality in 1943.

The flood of 1948, however, proved to be the much-needed spur for some action. This flood, brought on by a sudden spring thaw, culminated in a major catastrophe. It caused half-a-million dollars of damage and destroyed many downtown landmarks. Four years later a diversion channel was built. Constructed at a cost of $250,000, it was officially opened by Ontario Premier Leslie Frost on November 10, 1952.

The building of the channel was timely. On October 15, 1954, Southern

Nash was one of the companies that evolved into American Motors. Courtesy, Brampton Board of Trade

THE SHAPE OF GROWTH

Founded in 1908, the Hewetson Shoe Company was one of Brampton's oldest businesses until it closed its doors in 1979. The shoes in the advertisement, called "topsiders," have once again come into fashion. Courtesy, Brampton Board of Trade

Ontario was hit by a horrendous storm, Hurricane Hazel. Hurricane Hazel originated in Haiti, building up strength as it headed north, and when it reached Toronto it descended in a fury, lashing out with fierce winds and over nine inches of rain. Eighty-one people were killed, and 1,868 lost their homes. The water in Brampton's flood channel swelled to three times its normal capacity, lapping over the top, but it held. According to one newspaper report "the channel prevented a major catastrophe in Brampton."

The post-World War II period was one of rapid growth. The war had turned Canada into a force to be reckoned with and she emerged from the conflict as the world's fourth-largest industrial and trading nation. Attracted by the prospect of a prosperous future, thousands of immigrants, mostly from Europe, flooded into the country. Between 1941 and 1976, Canada's population jumped from 11.5 million to 22 million.

Brampton, too, was buoyant. Some 1,800 Bramptonians were still commuting to Malton. The aircraft industry thrived until 1959, when the production of the Avro was halted by the Diefenbaker Government. Hewetson's and Dale's were still major employers.

When Harry Dale started the business in the mid-1800s, he had one greenhouse warmed by a wood stove. If the weather was cold, somebody had

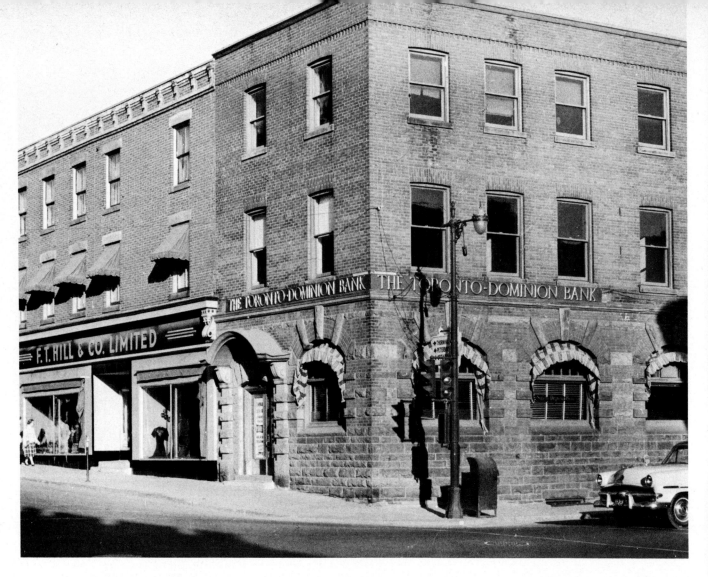

Brampton's beginnings can be traced to its "Four Corners," seen here in 1957. Courtesy, Brampton Public Library

Facing page: The Dominion Building clock tower suggests Brampton's historical past. Photo by Paul Sparrow

to feed the fire all night long to make sure the plants didn't die. A 1948 sales brochure indicates just how far Dale Estates had come since those early days. Thirty-five acres of glass covered 250-acres of gardens. There were 2,000 feet of underground tunnels. Dale's even had its own central heating plant with seven oil-fired boilers. The heating pipes stretched for a hundred miles, long enough, as one writer observed, to stretch two-and-a-half times between Toronto and Hamilton.

Meanwhile, many new companies came to Brampton—Moore Dry Kiln, Dixie Cup, Imperial Optical, Page Bros., Union Metal Company, and the Round Chain Company, to name a few. American firms, attracted by cheap industrial land and a willing labor force, opened branch plants. American Motors (Canada) Limited arrived in 1961. Shortly afterwards, the first Rambler manufactured in Canada rolled off the assembly line.

By 1959 Brampton had approximately fifty industries employing some 3,000 people. The value of industrial buildings increased dramatically. In 1940 such buildings were worth $10,600; by 1958 they were worth over $1.5 million. The population soared from 6,000 in 1949 to 14,500 in 1959—an increase of 124.2 percent.

By 1960 Brampton was on the threshold of a new era. Immigrants from Portugal, Italy, the Netherlands, Germany, and the British Isles (and more recently the Caribbean and Asia) altered Brampton's racial mix. No longer was the town predominantly British, in thought or in heritage. Brampton was changing. A city of diversity was emerging.

Left: Bramalea, planned as a distinct city, is now a part of the City of Brampton, and both are included in the Region of Peel, created in 1973. Photo by Paul Sparrow

The Civic Center. Photo by Paul Sparrow

Above: Waterskiing at Clairville Conservation Area on Brampton's east side. Courtesy, City of Brampton

Left: Meadowland Park is a favorite recreational spot. Photo by Paul Sparrow

Facing page, top: Snowmobiles bump over a snowy trail at White Spruce Park west of Heart Lake Road and north of Bovaird Drive. Courtesy, City of Brampton

Facing page, bottom: Sunshine and fresh air make for an invigorating game. Photo by Paul Sparrow

Left: Brampton's early days are reflected in the architecture of some of the buildings on Main Street. Photo by Paul Sparrow

Facing page: Brampton has many interesting eateries. This is the patio of Houston's at the Mill. The structure was a lumberyard in the 1800s and became a knitting mill at the century's turn. In the late 1960s the factory closed and a retail outlet was opened. Photo by Helga Loverseed

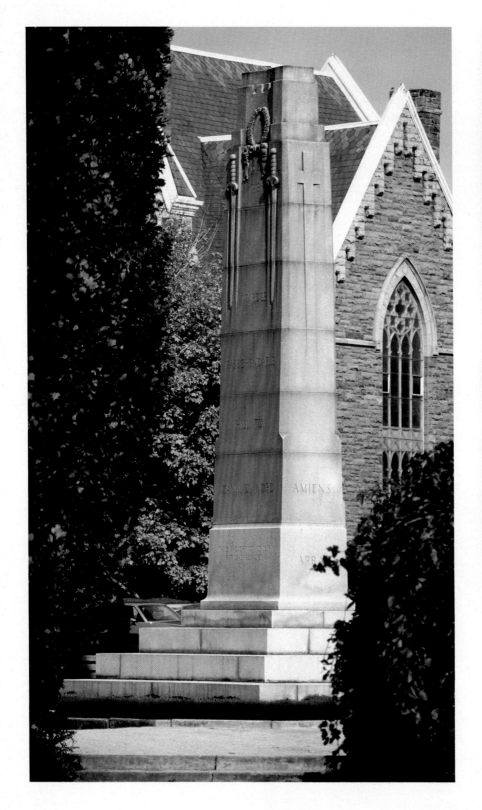

Left: A Brampton war memorial honors that community's citizens who defended Canada's and her allies' freedoms. Photo by Paul Sparrow

The old Peel County Courthouse (above and left) was completed in 1876 and, with the help of restoration work, is still in good condition today. It is a part of the Heritage Complex, and formerly housed a museum, which is in the Old Jail (next door) today. Photos by Paul Sparrow

Above: In the 1950s Brampton was best known as a centre for growing flowers. The city's slogan still boasts that "Brampton is Blooming." Courtesy, City of Brampton

Right: Fall harvest time still lends its particular beauty to the area that was originally known as quality farmland. Photo by Paul Sparrow

Gage Park (above and left) was named after Sir William Gage, a renowned Canadian publisher who was born near Brampton in 1849 and educated in the town. Top photo by Helga Loverseed; photo at left by Paul Sparrow

Above and right: Kids are the pride of any community. The child at right was the honorary "fire chief" at a Pine and Rose Parade, held on the first weekend in August. Top photo by Jac Jacobson; photo at right courtesy, City of Brampton

Above: Sheridan College. Photo by Paul Sparrow

Left: Central Peel Secondary School. Photo by Paul Sparrow

Facing page: A Brampton lake at sunset. Photo by Jac Jacobson

Above: Meadowland Park—a pleasant recreational area. Photo by Paul Sparrow

Conceived in the 1950s, the "self-sufficient" community of Bramalea was built to accommodate Toronto's burgeoning population with jobs, housing, schools, and shopping and recreational facilities. In early 1974 with the creation of the Region of Peel, Bramalea and Brampton, along with surrounding townships, were consolidated to centralize city services and streamline regional government, despite protests from some citizens. Courtesy, Lockwood Survey Corp.

CHAPTER SIX
CITY OF DIVERSITY

The last twenty-five years have, in some respects, echoed the development of Brampton in the mid-1800s. The city has changed enormously, of course, but there are similarities. Now, as then, Brampton is booming, and there have been tremendous social changes brought about by an explosion in the population.

The transportation system has improved in leaps and bounds. As the nineteenth-century railroads enabled Bramptonians to travel to Toronto in two hours, now modern highways allow them to drive to Metro in a mere thirty minutes. Changes in Brampton's legal status—first its incorporation as a village, then its incorporation as a town following the formation of the County of Peel —widely influenced Brampton's future. The creation of the Region of Peel and the consequent founding of the City of Brampton altered its destiny in 1974.

From the 1950s on, industrial development accelerated. Industries and subdivisions have grown up at a tremendous rate, so much so that the Brampton of yesteryear is practically unrecognizable.

Of all the developments that have helped to change the look and lifestyle of the town, perhaps the most dramatic was the birth of Bramalea. The brainchild of Bramalea Consolidated Limited (now the $2.2 billion Bramalea Limited), this was a man-made, self-sufficient "satellite city," created on what was formerly Chinguacousy farmland.

This development was built to cope with the overspill of population from Toronto. A 1957 brief to the Royal Commission on Canada's Economic Prospects had indicated that in order to support its burgeoning citizenry, Toronto would need an additional thirty-two square miles of industrial sites around or near the city. A self-contained community such as Bramalea, from whence people could commute, was seen as a way to provide the jobs and accommodations. Bramalea, however, was conceived as a separate city, not as a suburb of Toronto—a factor that was to cause considerable friction when Bramalea became part of the City of Brampton in 1974.

The name Bramalea was chosen because it was an amalgamation of Brampton and nearby Malton; Lea was an English word meaning meadow. Bramalea was, according to its creators, "descriptive of the location" and to some extent "descriptive of the satellite city-to-be."

Bramalea was a massive undertaking and it took years

CITY OF DIVERSITY

of research before it became a reality. In 1958, Close Brothers Limited of London, England, drew up an extensive master plan. The plan called for a development large enough to accommodate 50,000 to 90,000 people. It was to be "a marriage of ideas . . . between those of its creators (Bramalea Consolidated) and architect-planners and those of the provincial and township officials, business and cultural groups. . . ."

The idea was to create a harmonious blend of neighborhoods, industrial areas, and shopping and cultural facilities. There was to be "a distinct separation between areas for work, homes, and play." The land was to be split into residential areas (73 percent) and areas for business and industry (19.5 percent), with greenbelt in between—parks, golf courses and the like. In the City of Brampton today the land division runs around 68 percent residential to 32 percent industrial and commercial—probably a more realistic figure.

Some of the assertions made in that early master plan were idealistic, to say the least. The plan states, for example, that "traffic will flow smoothly and pedestrians will be able to move without fear of accident. Dust, dirt, noise and air pollution will not become problems. . . ." Of course, the planners of Bramalea were not to know that less than thirty years later, it would be part

Above: The planned community of Bramalea was planned to include residential and commercial areas, with parks, golf courses, and other recreational facilities available. Originally planned as a distinct city, Bramalea became a part of Brampton in 1974. Photo by Bill Huffman. Courtesy, the Bramp- ton Times *Collection, Region of Peel Archives*

Facing page: Terry Williams of Bramalea Secondary School in 1966 participated in a United Appeal relay race from the Bramalea Shopping Plaza to Roselea Park. He was congratulated on his victory by Brampton Mayor Russell Prouse and James Archdekin, Appeal chairman. Courtesy, Guardian *Collection, Region of Peel Archives*

163

of the City of Brampton—home to over 180,000 people. Unquestionably Brampton is a safe, clean city, but it is not without the headaches that a sizable population creates.

Bramalea was a massive undertaking. It called for the building of schools, offices, factories, homes, and stores. The land slated for development was a rectangular area of Chinguacousy Township. Unfortunately, parts were under the flight path for Toronto International Airport. Those areas were designated for industrial and commercial use, rather than for residential subdivisions.

Because it was such a large and wide-ranging complex, Bramalea was built in several stages. Subdivisions featured different styles of housing ranging from townshouses to high-rises to custom-built homes. They were built on spacious, landscaped lots, with winding, tree-lined crescents. A sales brochure described Bramalea as a "live-in, work-in, play-in city combining the best of both rural and urban worlds." Part of the marketing plan was to keep house prices at a moderate level. A family home could be purchased for between $20,000 to $30,000. The moderately-priced, modern housing soon drew people to the new community.

Meanwhile, Bramalea Consolidated was drumming up business for offices and industrial buildings. G.W. Finlay, then a vice-president of Bramalea Consolidated, went on an aggressive sales campaign, extolling the virtues of Bramalea throughout Canada and the United States. To induce companies to the area, industrial land was offered at give-away prices. Finlay's sales drive

CITY OF DIVERSITY

Above: By 1974, more than 100 new companies had moved to Bramalea, and American Motors was one of the first to arrive. Courtesy, Guardian Collection, Region of Peel Archives

Left: Bramalea's different subdivisions offered various styles of housing, from townhouses to highrises to custom-built homes. Courtesy, Guardian Collection, Region of Peel Archives

Facing page: The 1966 Junior Bramalea Lions Club lacrosse team. Courtesy, Guardian Collection, Region of Peel Archives

Above: The Bramalea City Centre, photographed in 1972. Courtesy, Brampton Public Library

Facing page: The ceremonial laying of the cornerstone of the Brampton Public Library by Dennis Timbrell and librarian David Skewe Melvin. This event occurred during Brampton's 1973 Centennial celebrations. Courtesy, Guardian Collection, Region of Peel Archives

CITY OF DIVERSITY

paid off. Northern Electric (Northern Telecom) and the Ford Motor Company were among the first companies to move to Brampton. They were soon joined by dozens of other large companies, including Kitchens of Sara Lee (Canada) Limited, Thomas J. Lipton Limited, Motorola Information Systems, and the giant, K-Mart. By 1974, over one hundred new companies had moved to Bramalea, providing work for 10,500 people.

By the middle of the 1960s, Bramalea had 1,500 homes. Industrial and commercial buildings occupied 3,100,000 square feet. The assessment of Chinguacousy Township had risen from $4 million (at the outset of development in 1959) to $24 million. Bramalea brought other benefits to Chinguacousy as well. The roads and fire service were improved and a police force was established. By 1966 the population had reached 15,996, twice as high as five years earlier.

Linchpin of the new development was the $20 million Bramalea City Centre. Today the complex encompasses stores, restaurants, movie theatres, and municipal offices for the Region of Peel. A $5 million Civic Centre was built next door. The Civic Centre houses Brampton's city council offices, the

BRAMPTON

mayor's office, an art gallery, a theatre, and a modern library. The City of Brampton now has three libraries: the Four Corners, Cyril Clark Branch (Heart Lake), and Chinguacousy (Bramalea). In addition to large collections of books, manuscripts, magazines, and other reference material, they are equipped with computers, microfilm machines, microfiche files and photocopiers—a far cry from Brampton's old Carnegie library.

While Bramalea was under construction, Brampton was doing some building of its own. Its population was rising steadily. In 1963 it stood at 26,363. Five years later it had climbed to 37,701. The town annexed farmland in order to build offices and factories and to accommodate housing for these extra people. Between 1946 and 1967, a total of 5,000 acres was acquired, extending the town boundaries and making Brampton twice the size it had been when incorporated as a town in 1873.

But shortage of land was not the only problem. The rising population was putting a strain on municipal services like the water and electricity. From the 1960s to the mid-1970s, the consumption of electricity had more than quadrupled, from a 1955 total of 30,052,118 kilowatt hours to 168,894,581 KWT purchased in 1966, and 301,809,375 KWH in 1972. To cope with the increased demand, Brampton Hydro-Electric Commission made a number of changes to the distribution system. Its capacity was increased and much of it moved underground.

In the 1950s, Brampton's water supply was still coming from underground wells and a couple of reservoirs, and by the 1960s they were woefully inadequate. An 830,000-gallon reservoir was built on Rutherford Road in

Above: American Motors (Canada) Ltd. is one of Brampton's largest manufacturers. These assembly workers are pictured in 1979 with the 25,000th Jeep to come off the line. Courtesy, Brampton Public Library

Left: To induce companies to the area, industrial land was offered at give-away prices. Northern Electric was among the first companies to move to Brampton. Today Northern Electric is known as Northern Telecom. Photo by Bill Huffman, courtesy, the Brampton Times *Collection, Region of Peel Archives*

1962. A well was also constructed at Huttonville. Residents, fearing that the well would damage the valuable farms and market gardens in the area, voiced their objections. Nevertheless, construction went ahead.

The capacity of the combined wells was supposed to be at least 5.8 million gallons per day, but between 1964 and 1965, water levels dropped by 50 percent—a potentially disastrous situation. Water rationing was introduced and building permits were suspended, as town officials tried to find an alternative source of water. It was obvious that Brampton now needed a pipeline to Lake Ontario. An agreement was struck with neighboring Toronto Township (Mississauga) through which the pipeline had to run. It was built at a cost of $500,000. Another $500,000 was spent on constructing a new storage reservoir.

The increased population was also putting a strain on the phone system—both in Brampton and in Chinguacousy Township. In anticipation of a higher demand for phones, the Chinguacousy Municipal Telephone system, a small privately run company, was absorbed by the Bell Telephone Company in 1858; as part of a large corporation it would be in a better position to service a fast-growing development like Bramalea. In 1974, Bell Canada invested $4.1 million on a new switching station, built by the Northern Electric Company (Northern Telecom) in Bramalea.

Brampton's telephone exchange—with a daily average of 26,000 calls—was one of the busiest in Southern Ontario, and many long distance calls were made to Toronto. Annoyed with the expenses incurred by long-distance dialling to Toronto, businesses started pressuring Bell Telephone to do something. This was done in 1971. Brampton became a part of the Toronto dialling area.

Many people who lived in Brampton commuted to Toronto to work because, despite the availability of jobs, the salaries in Brampton were not all that high. Dale Estates was notorious for its low wages (one reason why it employed so many immigrants) and even American Motors paid only $1.87 an hour to their assemblymen. Despite this unfavorable situation, few workers were members of a union. One reason seems to be that unions were frowned upon.

Terry Gorman, former president of Local 1285 of the United Auto Workers, recalls what labor relations were in the early 1960s: "In those days the union was looked on as suspect. This was more of a farming community then and people were conservative. . . . You could ask the boss for a holiday and he'd say there are many to replace you."

Inevitably, the workers soon began to favor the idea of joining a union. More and more of those workers were newcomers from other towns, accustomed to belonging to a body that could fight on their behalf. To coordinate the Brampton unions, an umbrella organization, the Brampton and District Labour Council, was formed in 1960. Peter McCombe, foreman at the *Brampton Times and Conservator*, was elected president. The Council's first members comprised the United Auto Workers; the International Typographers' Union; the Brampton Firefighters' Local; Sheet Metal Workers; International Tobacco Workers; and The United Glass and Ceramic Workers. Six years later, the membership had grown to include thirty-five local unions—a

CITY OF DIVERSITY

Brampton's excellent educational and recreational facilities have played a large part in attracting people to the city. Courtesy, Guardian *Collection, Region of Peel Archives*

sign of changing times.

Nowhere were these changing times more evident than in the field of education. The pattern of schooling set up by Egerton Ryerson in the mid-nineteenth century had developed through the years, and there had been an increasing emphasis on secondary education. In 1960, Central Peel Secondary School in Brampton set up the first completely reorganized program of studies in this part of Ontario (apart from Toronto). W.G. McDowell, the school's first principal, was instrumental in setting up the new courses. It was his firm belief that "98 percent of children are educable in varying degrees."

The courses were split into two-year, four-year, and five-year programs; many different options were offered. Four-year courses provided general education with some job training. Five-year courses were university oriented. They had a common core of compulsory subjects such as Business and Commerce; Arts and Science; Science, Technology and Trades—quite a change from the "three Rs" taught by Dame Wright, in Brampton's first school.

But one man, above all, helped to change the face of education, not just in Brampton but throughout Ontario—"local boy" William (Bill) Davis. Raised in "the Castle," a mansion built by George Wright, who brought the first railroad to Brampton, Davis is a member of Brampton's "Establishment." His mother was Vera Hewetson of Hewetson Shoes. His father was Grenville Davis, crown attorney for the County of Peel. Steeped in politics from an early age, Davis was elected to the Legislature to represent Peel County in 1959. Before becoming premier of the province in 1971 (a position he held for fourteen years), he served as Minister of University Affairs (1964-1971) and Leader of the Conservative Party.

Bill Davis's achievements as a politician are many, but some of his most lasting policies were made when he was Minister of Education (1962-1971).

Davis upgraded the school system and made it more efficient. This was quite an undertaking because it involved reorganizing school boards, closing schools that had outgrown their usefulness (at this time there were still many one-roomed schoolhouses in the province), and building new ones.

One of the first things Davis did was reduce the number of school boards from 3,676 to 1,673, a move that he achieved within five years. By 1969, he'd managed to reduce the number of boards even further, to only 192. At the same time, Davis authorized the building of hundreds of new schools. By 1965, 380 elementary schools and sixty-nine high schools had been constructed.

The most exciting, and certainly most controversial, change in the education system came about when Davis initiated the building of community colleges—Colleges of Applied Arts & Technology (CAATS). Essentially these colleges were an extension of the secondary school system, but they were designed to fulfill a role quite different from that of the universities. Not everybody was enthusiastic about the community colleges. In Claire Hoy's biography of the former premier, Bill Davis recalled: "We were criticized at the time for setting up a lesser system, but it wasn't that at all. There were

The city of Brampton today has three libraries. This is the interior of the Four Corners branch of the public library. Courtesy, Guardian *Collection, Region of Peel Archives*

Left: Premier Bill Davis cuts a ribbon to officially open the new GO-Transit service from Georgetown to Toronto. Courtesy, Daily Times Collection, Region of Peel Archives

Below: Brampton's Centennial was a widely celebrated affair. Events and activities involved all segments of the population and ranged from swim meets and soapbox derbies to historical walks and crafts displays. Courtesy, City of Brampton

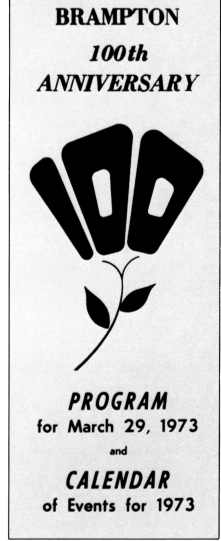

many reasons why the universities couldn't accommodate these students, and it's turned out to be a tremendous success."

In fact, the community colleges came along at just the right time. Canada was experiencing the much-touted postwar baby boom. There were more students than ever before, and the provincial government saw education as their ticket to success.

Premier John Robarts said: "Our true wealth resides in an educated citizenry; our shrewdest and most profitable investment rests in the education of our people. A general phenomenon of our day is that brain-workers are the prime economic need for societies in advanced stages of industrialism."

Scarborough's Centennial College (the first of twenty-two) opened in 1966. The following year Sheridan College of Applied Arts and Technology opened in Brampton. The college was built on the site of Brampton's old high school on Church Street and when it opened it had room for 400 students enrolled in twenty programs. Today the Brampton campus is just one of ten affiliated

BRAMPTON

CITY OF DIVERSITY

with Sheridan. The Brampton campus has moved to McLaughlin Road South (at Steeles) and it serves more than 2,500 students per year. There are dozens of full and part-time courses in a wide variety of disciplines—business, applied and visual arts, secretarial studies, computers, technical studies, and languages.

It was also during the postwar period that Brampton's Roman Catholic Separate School Board was created. Protestant and Roman Catholic children had attended the same schools—an unusual situation in a predominantly Protestant society—but the arrival of so many European immigrants, many of whom were Roman Catholic, pinpointed the need for separate education. Under the tutelage of Father C.W. Sullivan, priest of St. Mary's Parish, some 300 parishioners got together to form a board. Harry Kelley was appointed chairman.

The fledgling board had very little money. When the first school, St. Mary's, opened in 1957, students had to buy their own notebooks. Letters were delivered by hand, so that the Board could cut down on expenses. But as the Board became more established, it received needed funding. St. Mary's School was followed by St. Joseph's (1961), St. Frances Xavier (1965) and St. Anne's (1966). In 1969, the Brampton Separate School Board was absorbed into the Dufferin-Peel Separate School Board. Today with some 41,000 students, it is

Facing page: Susan Newmann, age twelve in 1973, helped to commemorate Brampton's Centennial. Courtesy, Guardian *Collection, Region of Peel Archives*

Below: Firemen recalled the early days of Brampton's fire department during the 1973 Centennial celebrations. In the old days it was customary to offer bonuses to the fastest "reel men." The Excelsior Hose Company #1, Brampton's first Volunteer Fire Service, was founded in 1883. Courtesy, Guardian *Collection, Region of Peel Archives*

Above: These girls rode on a vintage Chevrolet during Brampton's 1973 Flower Festival parade, which until 1974 was an annual event. Courtesy, Guardian *Collection, Region of Peel Archives*

one of the fastest-growing separate school boards in the country.

But if the beginning of the 1960s was a time of development and expansion, the beginning of the 1970s was an era of celebration and innovation. Between 1973 and 1974, the town celebrated its l00th birthday; Queen Elizabeth II and the Duke of Edinburgh paid a visit; GO-Transit, a fast commuter train, was introduced; and the Region of Peel (replacing the old County of Peel) was created, giving birth to the new City of Brampton.

These momentous events all happened within thirteen months of each other and during those months, Brampton, it seemed, was in a constant flurry of excitement. There were parties, parades, celebrities, and opening ceremonies by the score, as Bramptonians got together to celebrate one of their most memorable times.

To commemorate the centennial of Brampton's incorporation as a town, numerous social organizations and service clubs, not to mention dozens of

The 1973 Flower Festival parade included this float with a historical theme sponsored by Grace United Church—whose congregation in 1867 was the first in Brampton to construct a church building. Courtesy, Guardian *Collection, Region of Peel Archives*

volunteers, got together to plan special events, including the Brampton Downtown Promenade, parades of the Lorne Scots, a fall festival (Septemberfest), and a special "Brampton Day" at the Canadian National Exhibition. Commemorative dollar coins (redeemable only in Brampton!) were struck, and under the aegis of a Centennial Committee headed by Emmerson McKinney, a massive 100th birthday book, printed by the Charters Publishing Company, was assembled. Florence and Cecil Chinn, photographers in Brampton for many years, collated the illustrations and provided some of the pictures. C.V. Charters acted as chief historian.

Thursday, March 29, 1973, was a day of non-stop festivities. Opening centennial ceremonies were held at Gage Park. Children, let out of school at 2 p.m., skated and swam at Roselea Park. They stuffed themselves on hot dogs—over 6,000 according to a *Daily Times* report! In the evening there was another birthday party. Revellers gathered at Shoppers World, where "many

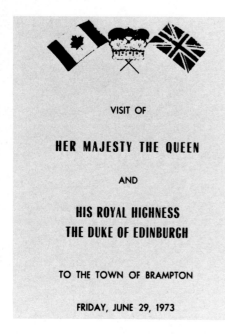

The highlight of Brampton's 1973 Centennial celebrations was a visit from the Queen of England and the Duke of Edinburgh. In Canada for a ten-day tour, they arrived in Brampton on June 29, 1973. Above, courtesy, City of Brampton. Right, courtesy, Guardian Collection, Region of Peel Archives

thousands danced and frolicked till the wee small hours."

Five thousand showed up for the opening ceremonies at Gage Park. The celebrations started in grand style. Bill Davis (then premier of Ontario), his wife, Kathleen, and Lieutenant-Governor Ross Macdonald and his daughter, Molly, arrived in a landau pulled by two fine hackney ponies. Lieutenant-Governor Macdonald then inspected an honor guard of the Lorne Scots, while planes of the 400th Squadron, Canadian Air Reserve, roared overhead. Meanwhile hundreds of children, skates still draped around their necks, arrived fresh from their party at Roselea Park. Waving flags and clutching balloons, they marched into the park led by the Senator Drum and Bugle Corps and Bobby Gimby dressed as the Pied Piper.

After the opening speeches, Master of Ceremonies John McDermid told the audience about his visit to Brampton, England. He had been warmly welcomed by its citizens and had brought back a framed copy of the original Brampton charter, granted to the Earl of Carlisle in 1604 by King James I of England. The charter was presented to Mayor James Archdekin who thanked the town for the "honor" they had granted him.

Premier Davis made a heartfelt speech in which he urged the crowd to "remember the men and women who had made Brampton what it is today." Referring to Brampton as "the best place in Ontario to live," he said: "We are in the process of change in this area and Brampton has always reacted positively to change and has always provided leadership."

The premier's speech gave a hint of things to come. Even as Brampton

CITY OF DIVERSITY

Mayor Jim Archdekin busily officiated many events during Brampton's 1973 Centennial celebrations. Courtesy, Guardian Collection, Region of Peel Archives

was celebrating its birthday, its days as a town were numbered. The following year it would become a city, part of the Region of Peel, making it much larger than before. In its centennial year Brampton's population was around 50,000—still, by some people's standards, a cosy little town. After annexation, it became a city with a population of 95,000, pushing Brampton into the big league.

The highlight of the centennial celebrations was a visit from Her Majesty, the Queen and His Royal Highness, the Duke of Edinburgh. In Canada for a ten-day tour, they arrived in Brampton on Friday, June 29, 1973. It was a marvellous opportunity for Bramptonians to enjoy some royal pomp and circumstance, and once again they flocked to Gage Park, as indeed they always did when they had something to celebrate. Thousands of locals and visitors turned up, excited by the prospect of catching a glimpse of the monarch. Her Majesty was met by Premier William Davis and Mrs. Davis and the Honourable Hugh Falkner, Secretary of State for Canada. The premier introduced the Queen to Brampton Mayor James (Jim) Archdekin and his wife. As Archdekin was to say later, it was the high point of his term in office.

179

BILL DAVIS

Ask any Bramptonian for one name of a hometown personality who made it to the top and the answer will probably be Bill Davis.

William Grenville Davis began his career as a shy, small-town lawyer with an eye on politics, and ended up at the helm of the Conservative Big Blue Machine and the Province of Ontario for a decade and a half. And through all of that, he remained just plain Bill.

It was from his father, Grenville, that Davis first caught the political bug. The senior Davis had always held a special feeling for Brampton, and he eventually moved to the town after growing up in Derry West. During Grenville's early years in Brampton, he focused on lacrosse. He was on the first Brampton Excelsiors team to make it to the national championships and he was eventually elected to the Canadian Lacrosse Hall of Fame.

The elder Davis graduated from Osgoode Hall in 1916 and then went on to be crown attorney for the County of Peel for thirty-one years. Because of this position, Grenville was unable to function in any political capacity. This did not stop him, however, from having many friends in politics, which was how the young Davis came to

William (Bill) Davis

obtain his unofficial training in politics.

Although Grenville was not active in politics, Bill Davis remembers him as being extremely knowledgeable in the political arena, and he would get into discussions with frequent visitors to his home, including MP Gordon Graydon, and Tory cabinet minister and former Ontario Premier Tom Kennedy. In fact, it was with Graydon that Bill Davis first attended a Tory national convention. At the age of sixteen, Davis was the youngest-ever convention delegate.

Davis's younger years were not all politics, however. Like his father, he had an aptitude for things athletic. Davis spent his childhood growing up at the home of his grandparents, the Hewetsons, because his parents moved there soon after their marriage. The house on Church Street was referred to by townspeople of the day as "the Castle." Davis spent hours on the home's expansive lot with neighborhood friends playing football. His love of the game remains today.

Initially, Davis wanted to enter federal politics and follow in the footsteps of Graydon, who was expected to retire. However, it was with the provincial Conservatives that Davis would make his mark. He contested the seat left vacant by Kennedy's death, using the political skills learned from Kennedy to win the nomination. By 1959 he was set to run against another Brampton native, Liberal William Brydon.

This first taste of elections was to prove a hard test for the rookie. Not long before, Prime Minister John Diefenbaker had put 14,000 aircraft workers, most of whom lived in Peel, out of work when he cancelled production of the Arrow fighter. Such a move did not endear Peel voters to a Tory, federal or not. Davis pulled out the victory by 1,200 votes, but in normal circumstances the margin should have been six times that.

Once Davis was in the legislature, his rise was rapid. Premier Robarts appointed Davis to his first cabinet in 1962 and named him minister of education. Two years later he added the title of Minister of University Affairs. He held both posts until 1971 when he ran for and won the party leadership. One month after winning the leadership, he was elected premier, a position he did not surrender until 1985.

Davis was constantly aware that his success in politics must have been very satisfying for his father, since as crown attorney he had not been allowed to enter the field. But Davis is quick to point out that there was never any pushing in that direction from his father.

Davis still lives in the home on Main Street that he and his father built over thirty years ago. This is not surprising, for Davis is as much a part of Brampton as anyone can get. Even as the leader of the province, he rarely failed to extol the virtues of Brampton no matter where he was speaking. Even now he speaks of Brampton's beauty and clean air when asked about his hometown.

Today, Davis is finding the time to do many of the pleasurable things for which there was never time before. As well, he still practices the law he learned as a graduate from Osgoode Hall. Whatever the future holds for Davis, there is little doubt that he will perform in a manner that Bramptonians have come to expect from their native son and most effective ambassador.

—Ken Moore

The Brampton Citizens Band played "Oh Canada." The Lorne Scots Military Band struck up "God Save the Queen." After speeches from the various dignitaries, the Queen rode in an open car up Main Street, waving to the many people who lined Brampton's sidewalks.

There were celebrations, too, when the GO (Government of Ontario) Transit train came to Brampton. GO-Trains had been running successfully on other routes since 1967. They provided a fast, efficient, moderately priced rail service for workers commuting to Toronto, and they cut down considerably on the hassle and expense of driving.

Brampton's service was inaugurated on April 27, 1974, when it made its first scheduled Georgetown-to-Toronto run. The timing was good, as Canada,

Premier Bill Davis at the controls of a GO-Transit train. In 1974 he took the train from Brampton to Bramalea to inaugurate the new service from Toronto to Georgetown. Courtesy, Daily Times *Collection, Region of Peel Archives*

like other western countries, was in the throes of the oil crisis. As usual Premier Davis was on hand to do the honors. Cheering crowds jammed Brampton's train station, and the premier, referring to Brampton's "historic day," had to yell to make himself heard.

As the crowd whistled and waved, two trains headed for Brampton station. The first to arrive was a sixty-two-year-old steam locomotive—a nod to Brampton's early trains. Then the GO-Train arrived. Without further ado, the premier donned an engineer's cap and drove the locomotive from Brampton to Bramalea. At Bramalea, former Chinguacousy Reeve Cyril Clark, clad as an Indian Chief in memory of the Township's original residents, demanded that Davis sign a "treaty" to verify that the train had passed through Chinguacousy territory.

Just as had been the case more than a century before, this train service was to provide a speedy and efficient link with Toronto. On its first run in 1974 it carried 1,000 passengers. The fare cost $1.20 (one way). Today a one-way ticket has risen to $3.30 (a monthly pass costs around $100). For many commuters the GO-Train is still more convenient than a car.

Brampton was on the threshold of a new era that was to lead to even more development and another rise in its population. The idea of introducing regional government—a more efficient form of administration than that which was in place—was once more a topic for discussion. The pros and cons of regional government had been discussed for several years, and in fact, a comprehensive study, called the Plunkett Report of 1965, had been put together after eighteen months of talks. In 1974 regional government finally became a reality.

Brampton's Go-Transit service to Toronto was inaugurated on April 27, 1974. For many communities the GO-Transit has proved to be faster and more efficient than automobile travel. Courtesy, City of Brampton

Centralized government in 1974 cleared the way for consolidation of public services. Neighboring police forces were amalgamated into the Peel Regional Police. Courtesy, Guardian Collection, Region of Peel Archives

By the end of 1973 the government had approved Bill 138, which created the Region of Peel. The Region comprised the newly incorporated, centrally located City of Brampton with the Town of Caledon to the north and the City of Mississauga, to the south. In addition to the "old" Brampton, the City of Brampton took in the townships of the Gore and Chinguacousy, which, of course, included Bramalea.

In a somewhat dramatic manner, the *Daily Times* of December 31 reported on the changes: "Brampton, Bramalea, Chinguacousy Township, Caledon Township and Albion Township—pretty well most of Peel County as we know it today—disappear at the stroke of midnight tonight . . ."

The idea behind regional government was to unify and centralize essential services such as water and electricity supply, education, road building and maintenance, and sewage disposal, and it was hoped to provide a more efficient form of government. By standardizing tax collection and distributing revenues evenly, wealthy areas could subsidize poorer ones. The Toronto Gore, for example, when it was an independent rural township, did not have a high enough tax base to support urban services.

Central government also cleared the way for the modernization of existing services. Brampton introduced a new bus system—Brampton Transit. In 1985, its fifty-five buses carried 4.7 million riders. Local fire deparments remained but the police forces were amalgamated. A central police station was built on Kennedy Road South.

Not surprisingly, many of these changes were met with opposition. The creation of the City of Brampton was a political hot potato, and Mayor Jim Archdekin didn't have an easy job promoting the idea. Those who lived in Bramalea complained bitterly. Many had moved to Bramalea precisely because it was built "from scratch." They liked the idea of living in a brand-new city, with an identity of its own. Now, many felt, Bramalea would lose that identity.

"Old Bramptonians" were also worried. They feared that "their" town would lose its distinctive personality. Gordon Beatty, interviewed in the centennial year, echoed the views of many old-timers. To him the "ideal Brampton" was that of the late 1920s, "still a small town but with many modern conveniences."

Despite the protests, the decision was made. On January l, 1974, the City of Brampton became a reality.

As time has proven, the positive factors soon outweighed the negative. Brampton is a mixture of the old and the new, a combination that to some makes it particularly appealing. T.J. Allen, director of the Communications Division of Sheridan College, commented in 1973: "It's partly because Brampton does have Old Bramptonians and a history as a county town, and has been prudent in making change, that the postwar migrants have been able to have a sense of identity with the town, of belonging to something real and stable, to a community."

Brampton still has a strong sense of place. The past has gone but it certainly has not been forgotten. The handsome nineteenth-century County Court building is part of a "Heritage Complex" which includes the Old Jail and former Registry Office, now home of the Region of Peel Museum and Archives. A few turn-of-the-century homes remain around the Four Corners area.

THE RICE GROUP

The Rice Construction Company was founded in Toronto by Will Rice and his three sons, Lou, Tom and Max. In the early 1950s, they decided on Brampton as their new base of operations. Some of the factors on which they based their decision to move to Brampton included the existence then of very few small home builders and the fact that no developments were being contemplated or were underway by a major developer; reasonable land costs; low tax rates; and a co-operative Town Council. In this period, there was a lot of movement of people and industry out of Toronto to the suburbs, and commuting from bedroom communities to industrial centres was gaining in popularity.

In 1953, the Rices acquired 300 acres of land from the Bull Estate and the Sterne and Armstrong properties all south of Clarence Street and east of Main Street. The name of this development, "Eldomar Heights," was derived from the given names of the wives of Will, Lou, Tom, and Max Rice. Completed by 1960, this project contained homes, apartment buildings, a shopping mall, and a number of industrial buildings. Homes were of high quality yet low-priced in the Eldomar development due to a unique method of component production and construction. In addition to being the first major developer in Brampton in 1969, Rice, in partnership with Kerbel Developments, built the first highrise apartment and seven stores on George Street. A large front page headline of the May 10, 1956, *Conservator* read "Local Building Boom Hits All Time High." Of $1,279,000 in building permits issued from January 1 to April 30 of that year, over $1,000,000 represented Rice Construction, as reported by building inspector Dennis Warren.

The Rice Group built more prestigious homes in a development known as "Ridgehill Manor." Construction got underway in 1960 with some participation by Syrnyk and Sons and Mike Jkachuk of Sonnet Developments Ltd. The company has more recently been jointly involved in some major Brampton developments on Highway 10 with Kerbel Developments.

The Rice Group can be credited with proposing a development policy which the Town Council adopted and made mandatory that the developer provide services, and that at least 30 percent of the development assessment was to be industrial. The requirement of industrial assessment did much to take Brampton from a bedroom community to a vibrant industrial and commercial centre. The success of Rices' developments did not escape the attention of other developers, with a consequent influx of many development firms in the years 1957 to 1960.

Notable of these early developers were the Kerbels, the Wingolds, Herb Green of Lloydtown Developments, and Charles Watson of Peel Village Developments.

In 1956 brothers Allan and Morris Kerbel commenced their first development comprised of 500 homes on 136 acres of what was the McLure Farm, known as Northwood Park. This proved to be the forerunner of further major Kerbel developments.

In 1958, an imaginative entrepreneur, Charles P. Watson, came to Brampton. With a background in insurance, mortgage financing, construction and developments, Charles Watson envisaged a completely new approach to large-scale land development. However, Brampton had all the requisites to meet the criteria for major development. Having decided on Brampton, Charles and his two associates formed Peel Village Developments Limited.

Their first acquisition was 600 acres of land bought from the Bull Estate. Further land acquisition from the Bull Estate and Armstrong Properties brought the landbank up to 700 acres for residential (5,000 homes), apartments and commercial development, and 400 acres for industrial developments. An additional 40-acre parcel was acquired west of Highway 10 north of Steeles Avenue, the site of Shoppers World Plaza and highrise apartments. May 1959 marked the beginning of residential construction which continues to the present day.

Brampton's picturesque Main Street still looks and feels like a small rural thoroughfare, and the Brampton Fair, an annual event since 1853, continues to delight Bramptonians of every age.

Looking at Brampton today, thirteen years after its birth as a city, it seems hard to imagine that any conflict occurred over incorporation. To be sure, Brampton has had its ups and downs. Like every other community, it suffered economically during the recession of the early 1980s. Several companies went under or sold out to other companies.

Hewetson Shoes closed its doors in 1979. Escalating leather costs and cheap imported shoes were blamed for its demise. Dale Estates, a victim of the energy squeeze, got out of the wholesale flower business. Its greenhouses became too expensive to heat. Moreover, jet aircraft started to bring freshly cut flowers from Spain and Holland in less than twenty-four hours. The blooms were sold at prices that undercut the home market and not even a company as large as Dale's could continue to compete.

New companies have quickly filled the economic void, and Brampton is developing at a mind-boggling rate. Barely a day goes by without a new building going up. Construction companies currently complete more than forty homes per week. Offices and factories are springing up all over the place. In 1985 alone (the most recent date for which figures are available) 206 acres of industrial and commercial land were developed. "Industrial development is very important," said Brampton Mayor Ken Whillans in 1986. "It provides jobs and keeps taxes low."

Brampton's population increases almost daily. In 1983, the population stood at 165,000. In 1985 it reached 180,000. Some sources predict that by the turn of the century, Brampton will be home to 300,000 people.

Pleasant living conditions and excellent recreational facilities have played their part in attracting people to the city. Brampton is, as Mayor Whillans said, "a people place with room to raise a family." In addition to 2,000 acres of parkland, there is a man-made ski hill, a 65-acre lake, and five golf courses. Indoor facilities include seventeen squash courts, thirteen fitness centres, nine arenas, a curling rink, seven indoor pools, two senior citizens' centres, and an indoor tennis centre with six courts.

Bramptonians today hail from every corner of the world, and Brampton has become a city of diversity. Its fifty churches represent many religious faiths and sects—from Judaism to Greek Orthodox. Even Brampton's annual festivals have adapted to the times. The Pine and Rose Festival (Pine for Chinguacousy; Rose for Brampton) grew out of the former Flower Festival and the Nitty Gritty Brama Ching Wing Ding. A new July festival, Carabram, was launched in 1982 as a celebration of Brampton's heterogeneous population. Pavilions representing many different cultures are manned by costumed hosts who entertain visitors and teach them something about their origins.

This multicultural mix is very much part of the City of Brampton, and if John Elliott and William Lawson were to return, they would surely be astounded.

But one aspect of Brampton hasn't changed at all—its entrepreneurial spirit. Last year Mayor Ken Whillans and a delegation of councillors travelled to Japan and the People's Republic of China. The purpose of the trip was to

CITY OF DIVERSITY

Brampton's Gage Park. Courtesy, Region of Peel Archives

promote Brampton and to persuade companies to invest. "Our basic message was 'Brampton is open for business,'" he said on his return.

Those consummate entrepreneurs—John Lynch, George Wright, John Haggert, Harry Dale, the Hewetsons, and all the others who got the ball rolling in the nineteenth century—would surely echo his sentiments.

Brampton's earliest settlers often had to hack their own roads through the thick forests, fording the Credit River and other streams in their wagons. Once at a homestead, the real pioneering began—clearing the land, building a cabin, and growing food for the family. Painting by C.A. Reid. Courtesy, Credit Valley Conservation Authority

CHAPTER SEVEN
PARTNERS IN PROGRESS

Members of the business community of the City of Brampton are convinced that Brampton is a city that cares about them. Whole planned satellite communities such as Bramalea offer a controlled environment where careful planning allows both industries and residential areas to flourish side by side. Industry gains a pool of workers close at hand, and residents gain easy access to a diversity of types of employment, in a city whose tax base is supported by business.

Many of Brampton's industries located in the city after considerable research into alternatives. In Brampton they found a somewhat unique combination of benefits. Canada's Golden Horseshoe is an area some forty miles deep, running roughly from the New York State border at Buffalo, around the west end of Lake Ontario, to Oshawa in the east. This area includes 25 percent of Canada's population, 25 percent of its wealth, and a large percentage of its industry. The entire area is within a relaxed, two-hour drive of Brampton. Highways 400 and 401 are within minutes of City Centre, providing truck access to all of Ontario from Windsor to the Quebec border and beyond.

Brampton's industries draw visitors from all across the world, while goods and technology travel to most world markets. The Toronto International Airport is within minutes of any part of Brampton, and Toronto's Great Lakes shipping port handles deep-water ships of all nations for eight months of the year. Rail service is good, with access to all of North America, and between the hydro power of Niagara Falls and the nuclear-generating capacity of Ontario Hydro, electricity is both plentiful and reasonably priced.

As modern industry continues to demand higher and higher technical skills from its work force, Brampton is within commuting distance of a vast pool of skilled, semiskilled, and relatively unskilled workers. Individuals requiring upgrading or improvement of skills have ready access to training facilities of all types, many sponsored or supported by Provincial or Federal Government training programs.

However, in the final analysis, what attracts people and industry to Brampton is a quality of life-style rarely equalled in North America.

The organizations whose stories are detailed on the following pages have chosen to support this important literary and civic project. They illustrate the variety of ways in which individuals and their businesses have contributed to the city's growth and development. The civic involvement of Brampton's businesses, institutions of learning, and local government, in co-operation with its citizens, has made the community an excellent place to live and work.

THE BRAMPTON BOARD OF TRADE

Built in 1906, the Carnegie Library building has housed the Brampton Board of Trade offices since July 1975.

The concept of boards of trade and chambers of commerce has been traced as far back as 2000 B.C. to the Mesopotamian City of Mari. In North America, the first was at Halifax in 1750, while the inception of the Brampton Board of Trade goes back 100 years to 1887. On June 14 of that year *The Conservator* reported that a meeting was held in the council chamber, with the mayor in the chair, attended by a fair number of businessmen from town. A board was formed from all present, plus those who had attended the meeting held two weeks earlier in the local Firemen's Hall. The first president was K. Chisholm and the first vice-president was A.F. Campbell, who, as editor of the newspaper, had advocated the creation of a board of trade.

During the first and second World Wars the Brampton Board of Trade became inactive; however, the present association has functioned continuously since being organized with the help of the Lion's Club in 1947. Its first office was opened in 1958 in space shared with the Brampton Real Estate Board on George Street. Two years later the organization hired a full-time secretary and moved to its own offices on Main Street South.

By September 1972 the board was able to hire Fred Hillhouse as full-time manager. In December 1973 it began to operate the Motor Vehicle Licence Bureau, originally located in the Bramalea City Centre and currently housed in the John Rhodes Centre on Airport Road.

Today the board has six full-time employees, under the direction of general manager Nancy Lumb, operating out of the old library building at 55 Queen Street East. President Dennis Cole has a board made up of 800 member firms, whose annual fees fund the operation. Volunteers provide the manpower for board projects, and in 1985 past president Robert Bell set up a system of 21 committees, each with responsibility for a specific project or function. A Governing Council controls the board and is comprised of 15 directors, 11 drawn from the general membership and 4 appointed. All are dedicated to ensuring that the Brampton Board of Trade actively represents and promotes the interests of the city's businesses and members, and indirectly, the interests of the people of Brampton dependent on those businesses.

Some of the projects supported by the Brampton Board of Trade include provision of regular awards for local business achievement, including Citizen of the Year, Business Person of the Year, and Farmer of the Year. The organization also makes policy recommendations to municipal and regional councils and initiates educational and social forums for its members.

CHEZ MARIE

In the 1930s Walter E. Calvert was one of Canada's leading rose growers and his home in this "Flower Town" of Brampton was surrounded by acres of greenhouses. Today his estate home stands in the midst of a modern subdivision in a rapidly growing city, yet within its walls the roses still bloom.

With the help of Marie and Joseph Colbacchin, who had a dream of transforming this grand old house into an elegant restaurant, the home was beautifully restored, renovated, and opened to the public in 1982 as Chez Marie. Located on Highway 10 just north of Brampton's Four Corners, Chez Marie was built on the reputation the Colbacchins had attained through their first restaurant venture in Brampton, Trattoria Via Veneto.

Joseph Colbacchin and his wife, Marie, opened Trattoria Via Veneto in 1978, serving a delightful blend of innovative northern Italian and continental dishes. Despite its unlikely location in a downtown office building, the cozy atmosphere, delicious cuisine, and the dedication and determination of the Colbacchins soon made Trattoria Via Veneto one of Brampton's most poular restaurants. The Colbacchins' reputation for gracious and professional service and excellence in fine dining preceded them to Chez Marie, where patrons are left with the impression of being a pampered guest in a grand old home of days gone by.

The house had many unique attributes such as the Belgian bevelled glass doors that were carefully restored and incorporated into the design of Chez Marie. Six elegant dining rooms have their own distinctive decor in garden tones of pink and green with artwork reflecting the historic rose theme of the home. A long-stemmed rose on each lace-covered table adds to the intimate ambience of Chez Marie. In addition to the superb French nouvelle cuisine, the Colbacchins have encouraged their talented chefs to feature wild game specialties such as venison, buffalo, yak, pheasant, and guinea fowl. It is this effort to provide patrons with unusual epicurean delights that has made a name for Chez Marie and Trattoria Via Veneto in Brampton.

Both restaurants have been reviewed in numerous Canadian publications, including the prestigious Oberon's *Where to Eat in Canada* dining guide, and have been honored with international recognition such as the Silver Spoon Award given by the Gourmet Diners Club of America. However, the greatest honor for Marie and Joseph Colbachin was in being recognized locally by the Brampton Board of Trade as Business Couple of the Year in 1984. It was with the valued patronage and encouragement from local residents that these two restaurants were able to "bloom" in the Brampton tradition.

Each intimate dining room at Chez Marie reflects its historic rose theme.

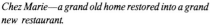

Chez Marie—a grand old home restored into a grand new restaurant.

BRAMPTON PLUMBING AND HEATING SUPPLIES (1981) LTD.

According to Wendell Brown, president of Brampton Plumbing and Heating Supplies (1981) Ltd., while the population of Brampton has grown dramatically in the last decades, the community still retains the ways and values of a smaller town. People are involved, the community encourages growth, and there is still a pride in personal and corporate loyalty. As a registered Bramptonian, born and bred in Brampton, Brown speaks from personal experience within the community.

When entering the firm's clean and bright customer showroom and offices, visitors notice the extensive rack of trophies that the company offers to local sporting events. Brown is directly involved in community sports, helping young people in hockey, lacrosse, and baseball. He is a member of the Brampton Rotary, and he and his staff are involved in United Way campaigns.

This sense of community involvement has contributed to the growth and evolution of Brampton Plumbing and Heating Supplies. In April 1962 three former employees of the Pease Foundry—Stan Shergold, Harry Ritchie, and Bob Seniscal—started the business with one employee, Wendell Brown. In August of that year the firm moved from its temporary offices, at McMurchy Street South and Sheard Avenue, to 275 Queen Street East. The company expanded to 18 employees, eventually outgrowing the facilities. In 1974 it moved to its present location at the corner of Orenda Road and Rutherford Road South, where the original 24,000 square feet of floor space has been expanded to 27,000 square feet by adding two mezzanine floors.

The last of the original partners retired in 1981, at which time the company was turned over to Richard Maltby, who became secretary/treasurer, and Brown, who became president. In September 1985 Maltby sold out, leaving Brown as sole owner. Brown's father worked with him until his death in 1972, and now his two sons provide the continuity of a third generation.

The firm's staff has grown from the original three partners and one employee to 42 employees, producing sales in excess of eight million dollars annually—split 60 percent industrial and 40 percent plumbing and heating. The delivery service, which is a trademark of the company, is provided by six vehicles.

In September 1984 Brampton Plumbing and Heating Supplies (1981) Ltd. purchased a small plumbing and heating supply company in Oakville to provide improved service to the Burlington and Oakville regions. The branch employs three people and operates one vehicle.

If you wonder how high a value Brampton Plumbing and Heating Supplies (1981) Ltd. places on loyalty to and from its customers, consider this. Wendell Brown still remembers that Fergus Wright of F.A. Wright & Son was the company's first customer.

The modern 27,000-square-foot office and warehouse on the corner of Orenda Road and Rutherford Road South in Brampton.

LUSTRO STEEL PRODUCTS

The Lustro staff in 1984.

Lustro Steel Products was purchased in 1962 by Joseph Flagg, founder of Essco Stamping in Windsor, Ontario. Flagg's son, Steve, was assigned to manage the firm until 1970, when it was sold to the Continental Group of Canada Ltd. Recently Continental, in turn, sold a 50-percent share to N.M.C. Canada, a division of Tang Industries of Chicago.

Lustro produces heavy automotive stampings in metal and is known for the quality and productivity of its organization. In 1984 it won *Canadian Machinery and Metalworking* magazine's annual competition for the National Productivity Award Certificate of Excellence for its design of statistical process control. According to general manager Doug Greer, Lustro is one of the first heavy-metal stampers to use the process.

The accelerating growth of Lustro Steel Products is evident in the size of its staff. In 1965 only 20 people handled the work. By 1982 there were 75, none of whom were laid off because of the serious depression in the automotive industry. In addition, three million dollars were invested to add 8,000 square feet and three new presses to the plant. That investment has paid off, so much so that Lustro now employs 150 people, 95 percent of whom are from the Brampton area. Relations between U.A.W. Local 1285 and the company are excellent, and work stoppages and strikes have been virtually non-existent.

Lustro Steel Products, at 40 Holtby Avenue in Brampton, was honored with an Outstanding Business Achievement Award of Merit from the Ontario Chamber of Commerce in 1986.

Lustro produces automotive stampings primarily for General Motors of Canada and Chrysler Corporation Canada. Each month the company ships 140,000 control arms, 125,000 transmission covers (all the transmission covers required for every Dodge Aries, Plymouth Reliant, and Chrysler Magic Wagon), and all the North American requirements for front bumper reinforcements on all Chevrolet Caprice models. With sales in excess of $32 million, Lustro Steel Products is one of the top automotive stampers in Canada, and in 1986 received an Outstanding Business Achievement Award of Merit from the Ontario Chamber of Commerce.

Greer believes the biggest change in the firm's business has been its customers' specified requirement for greatly improved quality. Buyers have lost autonomy to committees of all disciplines, and on a rating scale of one to 145, suppliers must attain averages of 142 to qualify. When Greer arrived at Lustro Steel Products, he observed one particular stamping being burred by less than perfect equipment. The customer accepted the product with a one-eighth-inch burr. Today that same defective part would be unceremoniously scrapped the minute it came off the line. Competition has made quality imperative.

ARCHDEKIN FUNERAL HOME (1985) LIMITED

Leo Archdekin, president and founder.

The Archdekin Funeral Home (1985) Limited was founded in 1965, when its president, Leo Archdekin, purchased the Cooper Estate home. The original historic old Victorian building has been maintained, much of it in its original condition. Needed additions were built on, first in 1974 when two sides and an expanded parking lot were added, and later in 1985 with the new 3,000-square-foot chapel, a minister's quarters, washrooms, and a special entrance with a wheelchair ramp.

Leo Archdekin retains autonomous control of the day-to-day operations, serving as president and director, to maintain the personal tradition and public acceptance of the funeral home.

Leo Archdekin is an Irishman with a ready smile and a comforting handshake. True to his farm upbringing, he still rises early to get the chores done and watch over his still-growing business. In its first year of business Archdekin Funeral Home provided only 36 services. By 1986 that number had grown to several hundred per year. The Archdekin Funeral Home is associated with worldwide affiliates to assist families at the time of death in foreign countries.

Archdekin believes in offering a total service to all denominations and faiths. He believes that when his customers are pleased, there is a 80-percent probability that he will be asked to serve any future needs of that satisfied customer. He is of the opinion that the funeral business is a personal one, best served by someone known in the community.

Archdekin is a member of the local Masonic Lodge, the Scottish Rite, the Peel Shrine Club, and the Brampton Curling Club. He is past charter president of The Canadian Red Cross Society, North Peel Branch, and has received its Distinguished Service Award. Archdekin also serves as president of the Metro Toronto and District Funeral Service Association, director of the Funeral Service Association of Canada, and is a member of the Guaranteed Funeral Deposits of Ontario. He has also been active in the local Lion's Club. One of Archdekin's most satisfying jobs is as "Ferdy" the Shriner clown in the Shriner Circus and in visits to hospitalized children.

Over the years there have been many changes in the funeral business. More people are using funeral homes instead of

The original Cooper Estate home.

The modern additions still incorporate the original Victorian home.

their own homes, and many are entrusting funeral directors with sufficient funds for prepaid funerals.

The largest funeral service ever handled by Archdekin Funeral Home (1985) Limited was that of Leo Archdekin's brother Jim, who died in office as mayor of Brampton. More than 3,000 people attended that service.

PARTNERS IN PROGRESS

M&P TOOL PRODUCTS LTD.

M&P Tool Products Ltd. is comprised of four separate working divisions. The Stamping Division produces light to medium metal stampings. The Space Aid Division provides a variety of material storage and handling equipment, while the Water Bed Division manufactures water bed care products, hardware kits, software kits, and furniture hardware.

The company was founded in 1968 as a tool and die shop, with president Max Prufer as its sole employee. For the first six months the firm was located on a farm in the Hockley Valley near Palgrave. By 1969 it had expanded to an old airport hangar opposite the fair grounds on McMurchy Avenue South.

Three years later M&P moved into a 5,000-square-foot plant on Stafford Drive. Despite an addition of 5,000 square feet, the plant was soon outgrown. In 1978 the company moved to its present stamping plant on Advance Boulevard.

At that time Pedlar Engineering was a customer, making storage racking nearby. When it closed, M&P hired the fabricating crew and its foreman, and entered the specialized storage rack business.

The management team was reorganized and restructured in 1981. Max Prufer remained as president, with his son Jim Prufer as vice-president. Manfred Meyer, the foreman hired from Pedlar, is now manager of the fabricating plant on Strathearn, while Martin Prufer is manager of the plant on Advance Boulevard. The final member of the team is Mrs. Grainger, a 12-year employee who is now secretary/treasurer. Each of these people has common stock and a say in the future of the business. The change has paid off; sales at M&P Tool Products have increased from one million dollars in 1981 to a projected five million dollars in 1986.

Max Prufer started work as an employee in his father's bicycle shop in East Germany. He is now a skilled tool and die maker and designer. His company currently has accounting functions on computer and is moving into hi-tech industrial computers for the factories.

The M&P management team is skilled in all areas pertaining to the firm's enterprises. By offering the capability to design custom shelving, racking, and individually required stacking containers, while also being flexible in its approach to novel customer requests, M&P Tool Products Ltd. will continue to attract projects. For example, in 1986 it received the contract for the bulk of the material handling equipment for the new Honda plant.

Max Prufer, president, who founded the original tool and die shop in 1968.

M&P Tool Products Ltd.'s facility on Advance Boulevard.

COLONY LINCOLN MERCURY SALES LIMITED

Colony Lincoln Mercury Sales Limited consistently ranks among the top dealers in the Toronto area in most categories of sales volume and customer satisfaction, and in 1968 was named one of the top 40 dealers.

The dealership opened in Brampton in 1967. Four years later its current president, Keith Coulter, purchased the organization with the financial assistance of the Ford

ABOVE: Inside the Colony showroom.

LEFT: The Colony offices and showroom viewed from Queen Street.

Dealer Development Plan. At that time the dealership had 40 employees producing a sales volume of approximately three million dollars. Today the firm employs more than 60 people, an increase of 50 percent. Its sales volume has increased to $30 million, representing a jump of 1,000 percent.

Many years before coming to Brampton Coulter was asked, given his choice of anywhere in Canada, where would he most like to establish a Ford dealership. Recognizing the growth potential of the city, he chose Brampton, and considers himself most fortunate to be here now. "If you aren't doing well in Brampton, you won't do well anywhere," he says.

However, the growth of Colony Lincoln Mercury Sales Limited is more than just a simple correlation with the growth of Brampton. The organization approaches business with some old-fashioned but effective concepts. The firm is blessed with a stable total staff, and, more importantly, a stable management staff. Morale is important at Colony, and, where internal advancement is blocked for a deserving employee, the firm supports the effort to find that employee a position elsewhere that offers advancement. To

Good housekeeping, even in the service bays, reflects excellence in company morale.

date, two former employees have been assisted in progressing to ownership of their own dealerships, and various other positions have been arranged and are being arranged.

In addition, Colony has made a major direction change, becoming heavily involved in sales of heavy trucks and the daily rental business.

Coulter is a past president of the Brampton Board of Trade and a longtime Rotarian. He and his staff are involved with fund raising for various projects including the local Salvation Army. In addition, the firm supported a woman in her efforts to try for Canada's Equestrian Team.

Colony's contribution to the community includes support of local baseball, soccer, and hockey teams for both young men and young women. This is in keeping with other old-fashioned concepts. The organization stresses that it is in business to show a profit, but it won't be at someone's expense. Colony Lincoln Mercury Sales Limited is in business for the long term, and that approach shows up in staff morale, in the cleanliness of its facilities, and in the way it operates.

PARTNERS IN PROGRESS

ARMSTRONG HOLDINGS (BRAMPTON) LIMITED

Armstrong Holdings (Brampton) Limited was established in the late 1960s, undertaking the responsibility for what had been the Armstrong Brothers Company's A.B.C. Farms Limited. The construction end of the business was spun out under the Armbro Holdings banner.

At 14 years of age chairman Charles Armstrong's father, Elgin, took over the family farm, and with his brother, Ted, built the Armstrong Brothers construction and farming giant. Their story starts and continues with winning.

Elgin Armstrong was a prize winner for his alfalfa seed at the first Royal Agricultural Winter Fair. The hackney pony Crystal Lady won the Royal Winter Fair seven years in a row, and her granddaughter has won four years in a row. The A.B.C.-bred Holstein-Friesian bull 7 A.B.C. Reflection Sovereign was the greatest in the world. Over the years there have been prize-winning hackney ponies, Thoroughbred horses, and jumpers, but mostly there are the standard-bred horses. Today Armstrong Holdings concentrates only on the breeding, sales, and racing of standard-bred trotters and pacers.

The prestigious Hambletonian Cup was brought home to Canada for the first time in 1953 when the Armstrong's two-year-old-filly Helicopter won an upset victory. Shown as they accepted the award are Mrs. J. Elgin Armstrong (centre), C. Edwin Armstrong (second from left), and J. Elgin Armstrong (far right).

It all started in 1952, when Elgin Armstrong bought the two-year-old standard-bred filly Helicopter, and won an upset victory at the prestigious Hambletonian Stakes at Goshen, New York, the following year. Since then Armbro horses have won all over the world. North American Classic races won include The Adios, American Classic, Canadian Pacing Derby, Cane Pace, Dexter Cup, International Stallion, Kentucky Futurity, Little Brown Jug, Maple Leaf Trotting Classic, Messenger, Prix D'Ete, Roosevelt Inter-

More than 400 horses are bred each year at the 1,200-acre, farm-complex headquarters of Armstrong Holdings (Brampton) Limited at Victoria, just north of Brampton.

national, and Shapiro. The newest star is home-bred filly Armbro Fling, who has already won the Acorn Stakes at Meadowland, the elimination to enter the Merrieannabelle, as well as the $600,000 Merrieannabelle, coming from sixth in the stretch.

The corporate headquarters of Armstrong Holdings (Brampton) Limited is the new farm complex at Victoria, just north of Brampton. The 1,200 acres are all safety-enclosed with chain-link fencing and automatic opening and closing gates. Aside from the office and eight barns, the entire property is in pasture. Each Armbro orange barn is 250 feet long, houses 100 horses, and is set back from Highway 10 in stages so they can all be seen from the office. The staff of 30 to 50 people includes president Dr. J. Glen Brown, who has been with the Armstrongs for 26 years, and two other full-time veterinarians. More than 400 horses are bred at the facility each year. The yearlings are either sold in Toronto or Lexington, Kentucky.

E.A. MITCHELL LIMITED REALTOR

A modest beginning, the Mitchell office in 1957.

In August 1957 E.A. Mitchell founded and opened Mitchell Real Estate with one employee. Today the firm employs more than 70 people and returns more than $4.25 million in salaries to the community where it does business.

That first employee was Joe Harley, now a major shareholder and director of the company that incorporated in 1960. Other shareholder-directors are E.A. Mitchell, president; Donald Haynes, who joined the firm in 1959; and Carl Markvorsen, who also started with the company in 1959. All directors have completed the three-year real estate brokerage course given by the University of Toronto, and each has been made a Fellow of the Real Estate Institute. In addition, all four continued studies at Ryerson Polytechnical Institute. Mitchell and Haynes studied appraising, while Harley and Markvorsen added property management to their list of skills. Today Harley is an acknowledged expert in property management, having taught courses at Sheridan College and other national and provincial institutions. In addition, Markvorsen manages the largest industrial sales division in Brampton.

The Residential Division of Mitchell Real Estate handles the sale of houses, apartments, and condominiums, producing 55 percent of the business. The I.C.I. Division (industrial/commercial/investment) adds 30 percent of the firm's revenue. Investment is in plazas, apartments, and factories purchased to provide an income source. The industrial and commercial sides sell development land, stores, and offices. The balance of E.A. Mitchell Limited's business comes from appraisals.

Mitchell notes his firm does not manufacture, and thus its only product is service. To sell this concept, he has attracted people who believe in being good neighbors as well as good business people. All four directors have served as president of the Brampton Real Estate Board, and Harley is the founding president of the Downtown Businessmen's Association. Both Mitchell and Harley have served as president of the Brampton Board of Trade. Haynes serves with both his local church and the Salvation Army. Markvorsen is involved with the Rotary Club, participating in its fund-raising drives. Mitchell is a past president of the Ontario Real Estate Association, a former vice-president of the Canadian Real Estate Association, and has been chairman of the United Way and the North Peel section of the Cancer Society fund drive as well as chairman of Peel Memorial Hospital expansion campaign. A goodly number of the employees of E.A. Mitchell Limited are also very active volunteers throughout many varied community endeavors.

The community has noticed: Joe Harley was voted Businessman of the Year in 1983, and Mitchell was voted Citizen of the Year for Brampton in 1984.

The four shareholder-directors of E.A. Mitchell today are (left to right) Don Haynes, Joe Harley, Earnie Mitchell, and Carl Markvorsen.

PARTNERS IN PROGRESS

BURLINGTON CANADA INC.

The Burlington Canada Inc. office and manufacturing facility—over 10 acres under one roof.

Burlington Canada Inc. is one of Canada's largest manufacturers of carpeting with close to 700 employees. Burlington first came to Canada in 1967, employing only 50 people. Today the firm occupies a 475,000-square-foot manufacturing plant and a 125,000-square-foot distribution centre in the City of Brampton.

Burlington Canada is a subsidiary of Burlington Industries, the largest textile manufacturer in the world.

The rapid growth in Burlington Canada is a result of leadership in research and development, resulting in improved products at reduced cost. The company is the fashion leader of the Canadian carpet industry.

Burlington is very community conscious and is proud of its multicultural workforce, particularly in the case of Lisa Buscombe, an employee who recently won the Provincial, national, and, finally, the title of Women's World Freestyle Champion Archer at the competition held in Hyvinkaa, Finland.

Burlington Canada Inc. and its people—contributing to their community in a variety of ways.

NATIONAL GENERAL FILTER PRODUCTS LTD.

National General Filter Products Ltd. manufactures industrial filters for installation in the air-filtration systems of large commercial buildings and industrial spray paint booths.

To comprehend today's industrial filter market, it helps to know some history. As late as 1960 practically all commercial buildings filtered their incoming air through a system of washable metal screen air filters. The filters were oil spray coated and installed in racks in front of the heating or cooling

Modern society generates industrial by-products that pollute our cities and work places. National General Filter Products Ltd. is a leader in researching ways and means of making the inner environment cleaner and safer.

coils. Fans behind the coil wall drew the air through the coils, heating or cooling the air depending on the season. This incoming air contained dirt from the recirculating air in the building as well as the outside airborne contaminants. The oil-soaked air filters trapped some of this dirt as it found its way through the mesh layers and eventually plugged up, cutting off the air flow. The oil also found its way down stream, plugging up the coils and fans as well as leaving telltale messes on walls and ceilings. Cleaning of these filters was extremely messy as the dirty, oily panels had to be removed to another area sometimes down the elevator to the basement leaving a trail of oil in their travel. The filters were cleaned with a hand-held steam cleaner and then emersed in a tray of oil. The maintenance people then had to retrace their travel back up to the penthouse area on the roof and replace the filters into their racks. Not only was this messy, but the labor costs were starting to add up due to the fact that this work was usually done in the evenings or weekends.

In the early 1960s atmospheric dust conditions around the large cities were worsening with the increase in incinerators, automobiles, and industrialization. The dirt entering commercial buildings was rapidly becoming more of a problem and created a need for more sophistication in air filtration.

National General Filter Products Ltd. was created out of a strange set of circumstances originating with Johnson & Johnson (the baby powder people) hiring Alfred English to work in its design department in the fall of 1963.

English was to work on the design of air-filtration systems using a new approach called blanket filtration. This new product incorporated the use of chemically impregnated polyester material that could be used in large panels usually 40 inches wide by up to 12 feet high. This material was held in the frames by a plastic retaining extrusion. The material was rolled into the extrusion using a roller tool similar to a pizza cutter. This system replaced the messy, less-efficient washable filtration systems and allowed for easy replacement. When dirty, the pad was simply rolled up from off the gridwork holding the dirt inside the roll.

In 1964 General Motors converted all its spray booth filtration systems in Oshawa and used the new filter in its new St. Therese, Quebec, plant. Ford was building a new plant in Oakville and American Motors was establishing a facility in Brampton, and both companies included this new approach to filtration in their spray booth design. A number of commercial buildings in downtown Toronto replaced their washable systems with blanket filtration, and the move was on to expand the blanket filtration concept to all types of air-filtration applications.

During this period of rapid acceptance of blanket filtration, English was instrumental in working closely with General Motors in the filtration design for its St. Therese plant, but had experienced some problems with the plastic extrusion that held the filter media in place melting and allowing dirt to bypass the filtering material.

In 1965 English left Johnson & Johnson and went on to a sales position unrelated to air filtration until one day months later the solution to General Motor's filtration problem suddenly came to him.

The retainer should be made from a formed metal shape rather than extruded plastic! Several shapes were tried and, when the right one appeared to be the answer, English took his idea to the General Motors engineers and presented them with his solution to their problem.

The product received immediate approval. English was presented to

A blanket filter in an industrial application. Air is being drawn through the filter to remove dirt before the air enters and cools an electrical switch room.

purchasing, and an order was placed for approximately $65,000 worth of Filter-Lok retainers. The purchasing agent, of course, asked English the name of his new company (it wasn't formed yet), and English, scrambling in his mind for a company name replied, "General Filters."

English was now in the air filter business, and a long road lay ahead. During incorporation it was found that General Filters was already incorporated as Canadian General Filters (oil line filters), so the name National General Filter Products Ltd. was chosen.

National General Filter Products Ltd. soon set up a work area in the back half of a small plant in Etobicoke, and in the first two years developed a rather crude blanket media of its own. This media was derived from a quilting material manufactured by Jaro Manufacturing in Woodstock, Ontario. The rolls were shipped into the factory and rolled out on the floor approximately 90 feet long. A mixture of contact glue, mineral oil, and orange dye was heated, using a rented gas burner, then poured hot into a rented spray gun for spraying onto the material before it cooled. It wasn't long before this activity wore out English's welcome, and a new location had to be found.

National moved to Brampton in 1967 and re-located in an old auto body shop behind Parkinson Bus Lines on Kennedy Road, north of Queen Street. The body shop had a separate room that previously had been used as a spray booth for painting cars and was now used for a chemical spray area needed to produce National's main product, Polytac blanket filter media.

The other room became an inventory storage area and an even smaller room became the office. The whole factory size had grown to less than 1,000 square feet, and National increased its employee base from two to three with the addition of a secretary.

The firm soon found itself with products ideally suited for an industry in its infancy. Within two years of working out of the body shop location, National moved to a small factory on Research Road, but again soon ran out of space. The next and last move was to the company's present location, at 100 Rutherford Road South, in the fall of 1971.

National's management didn't know at the time that Stearns & Foster had moved in next door in the same building and had the capacity to manufacture the raw material used in National's product. This fact was unknown until 1978, when Jaro Manufacturing, National's original supplier, sold out to Dominion Textiles and stopped manufacturing polyester filter material.

The timing couldn't have been better; National was released from its obligations to Jaro Manufacturing and immediately struck an arrangement with Stearns & Foster to produce its material requirements on an exclusive basis.

The two companies have worked well together ever since, with a conveyor system in place between Stearns & Foster's production line and National's manufacturing operation. Stearns & Foster soon became one of National's main suppliers and National, in turn, one of Stearn's larger customers.

In 1977 National General Filter Products Ltd. bought out one of its larger competitors, Facetelle Enterprises, formerly owned by Fram Corp., and started to add new products to better compete in this rapidly growing industry. National General Filter Products Ltd. was built on the development of new products all related to improvement of air quality in the working environment, as well as the cleaning of air exiting out of manufacturing operations before it becomes an irritant to the neighborhood.

In the 20 years that National General Filter Products Ltd. has been located in Brampton, the company has been fortunate to attract local talent and put together a strong professional manufacturing and management group. Some of its employees have been with the firm for more than 15 years and have become an important factor in its growth.

The company has expanded over the past 10 years to become a national operation with sales offices and factories located in Vancouver, Calgary, Edmonton, Saskatoon, Winnipeg, London, and Montreal. This expansion has made it possible for National to compete favorably against some of the larger international air filter companies for a larger market share.

The first 20 years of new product development and marketing strategy have established National General Filter Products Ltd. as a leader in its field in Canada. The company's future plans include a modernization and expansion of its research and product development capacity to better launch it into the international market.

FBM DISTILLERY CO. LTD.

The initials "FBM" stand for Facundo Bacardi y Maso, second son of Don Facundo Bacardi y Moreau, the Catalonian emigrant who founded the Bacardi rum organization in 1862. On February 4 of that year he purchased the primitive Nunes distillery for $3,500. In 1960, when Fidel Castro nationalized the Cuban portion of the firm, it was valued at $76 million. The intervening years were not noted for their placid political nature.

The year 1868 heralded the start of the Ten Years War as Cuba sought independence from 400 years of Spanish domination. Don Facundo's son Emilio worked for the underground; he later was caught and exiled for four years on a Mediterranean island off Spanish Morocco. In 1895 Cuban patriot Jose Marti invaded, and the Cuban people rose in rebellion. This time Emilio was exiled to Morocco. The U.S. battleship *Maine* was sunk in Havana harbor in February 1898, and by August Cuba was independent. Emilio came home a hero. He went on to become the first freely elected mayor of Santiago and later was elected a senator.

However, war is not all hell. A U.S. Army officer invented the Bacardi rum Cuba Libre in a Havana bar when he mixed the rum with Coca-Cola. And it was a U.S. engineer who invented the Bacardi rum Daiquiri by mixing white rum with icy lime juice.

Bacardi founder Don Facundo was a wine merchant so trusted in Cuba that he was put in charge of food relief following the great earthquake in 1852. Before 1862 rum was often a thick, dark drink with a pungent taste. Don Facundo perfected the distilling and blending process and was able to use molasses to provide a light and mellow rum. By 1876 Bacardi rum had won its first international gold medal from the U.S. Centennial Exposition in Philadelphia. Among other awards were gold medals from Madrid in 1877, Matanzas (1881), Barcelona (1888), Paris (1889 and 1900), Chicago (1893), and Bordeaux (1895).

In 1910 the company proved that the Bacardi expertise could travel, expanding to Barcelona. New York was added in 1918, Mexico in 1931, Puerto Rico in 1935, and Canada in 1969. After the Castro takeover, the head office was moved to the Bahamas. Today Bacardi rum is distilled in those countries in addition to Spain, Brazil, Martinique, Panama, and Trinidad. Additional bottling plants are located in Australia, Austria, France, New Zealand, Switzerland, the United Kingdom, and West Germany. Bacardi continues to sell the largest volume of rum in the world.

The secret of Bacardi rum is passed from generation to generation and is retained within the Bacardi family. Roberto del Rosal, chief executive officer of FBM Distillery in Brampton, is a great-grandson of founder Don Facundo. The successful preparation of Bacardi rum starts and finishes in the laboratory. Today that process begins in laboratories in Nassau, Bahamas, and Brazil where the fermentation is done. Under controlled conditions, diluted molasses is placed into a sterile test tube and a tiny pinpoint of yeast is added. The yeast acts on the molasses, and its cells multiply, converting the sugar to alcohol. Over the next week to 10 days, as the yeast grows, it is transferred to larger and larger containers, until the fermenting culture is large and active enough to resist outside contamination. Then it is moved to vast stainless-steel tanks where the fermentation is completed in 72 hours.

Twice a year Bacardi International Ltd.'s own ship brings 600,000 gallons of fermented liquid to Canada. When FBM settled in Brampton, a deciding factor was the availability of two million gallons of water per day, which is what would be needed if the fermentation process were done there. FBM introduces the fermented liquid into a continuous-column still, then draws off the pure rum distillate using live steam to ensure a very high proof, which eliminates impurities and harshness.

Next the distillate is fed through stainless-steel tubes to the American white oak barrels for mellowing in wood. Bacardi light rum comes from natural wood barrels. Those barrels will last for up to 40 years, but for reuse need the inside to be charred, yielding a dark rum with its distinctive color and flavor.

The final step is the combining of rums from different batches and ages according to the Bacardi family formula, with any balance adjustments dictated by the trained nose and palate of an expert rummaker. Results are confirmed acceptable first by the local lab, and finally at the head office before shipment is allowed. The end product is always uniform and popular.

When Alberto Bacardi, the former president of FBM, surveyed the Canadian market in 1964, the firm sold 120,000 cases per year. He started in Canada with five employees in 1968; that number grew to 21 the following year when the first plant opened. In 1975 the new facility was completed, and the office that had been downtown was combined with the plant. Today FBM Distillery Co. Ltd. provides more than 75 jobs in a bright, airy, immaculate, and remarkably well-maintained facility. With sales of 1.1 million cases per year, Bacardi rum is the number one selling brand in the country—and the Bacardi bat has found a home in Brampton.

PARTNERS IN PROGRESS

FBM Distillery Co. Ltd.'s history goes back to its founding in 1862 by Facundo Bacardi y Maso on the island of Cuba. Today its Canadian home is in Brampton.

ATTRELL TOYOTA

Guild Toyota was founded by Norman Guild in 1967. From the Gulf service station on the corner of Kennedy Road and Clarence, he expanded into the unit next door with a new Toyota franchise. When road work by the City of Brampton cut off access, the firm moved to Queen Street East, where the Ali-Baba Steak House, which is now Krebs, is located today.

Robert B. Attrell, president, bought the firm in 1975 when Guild decided to reduce his involvement. Attrell had been with Toyota Canada Inc. and, despite the decline in imported car sales between 1972 and 1975, recognized the potential in Brampton. He first trimmed the operation to ride out the slow times, moving the showroom to Main Street South, the office to a trailer, and the parts and service departments to Clarence and Selby. In August 1976 he moved the business to 250 Queen Street West, and five years later moved the firm to its present location at 243 Queen Street East.

Known as Guild Toyota and headquartered at the 243 Queen Street location (above) since 1981, the agency will soon occupy the Consumers Gas Building (left) and become Attrell Toyota.

When Attrell started with the business in 1975, Guild Toyota had a one-car showroom, three service bays, a staff of seven, and sold 124 units. At its present location a staff of 20 sold more than 500 units in 1986. Attrell Toyota will replace Guild Toyota when it moves to new premises in what is now the Consumers Gas Building. The $2-million facility will double the firm's existing space, triple the frontage, and provide two entrances. Sales of 1,000 units are projected for 1990.

Attrell is a member of the Bramalea Baptist Church and over the years has worked with the Rotary, Lions, and Kiwanis. He is a member of the Brampton Automobile Dealers Association. Jointly and on his own, Attrell sponsors hockey, soccer, baseball, and basketball teams in the community. But when you see the gleaming old yellow Lotus racing car in his showroom, you begin to suspect his real involvement is with cars.

Attrell serves as chairman of the Toyota Dealer Council, and is past president, member, and active spokesman for the Association of Japanese Auto Dealers. In addition, he is a past president and member of the British Automobile Racing Club. For many years Attrell actively raced and only recently retired from 20 years of racing. His club B.A.R.C.oc was chosen to organize Toronto's first Indy-type race at Exhibition Place and he had the honor of driving the pace car.

Sally McCoy is office manager, bookkeeper, and secretary; she has been with Guild almost since its inception. Continuity is also assured through Bob Attrell's sons' involvement in the business. The eldest, John, is currently sales manager and is expected to be general manager. David is learning the service end of the business, while Robert Jr. is in sales. The younger boys will probably branch out on their own eventually, but for now Attrell Toyota boasts a motivated and dedicated staff.

Robert Attrell (right) and son, John, with the Missionary Van donated by Guild Toyota.

ARMBRO HOLDINGS LTD.

The Armbro Bros. Construction was founded in 1929 by two brothers, J. Elgin Armstrong and C. Edwin Armstrong, who entered the business using their farm wagons to haul gravel and water for road work. By the mid-1930s Armstrong Bros. Construction had finished its first highway project complete with ditches, culverts, roadbed, and gravel surfaces. Since then the firm has developed into one of the

In the good old days of steam power.

largest civil engineering contractors in Canada, completing such diverse projects as the runways at Mirabel International Airport, subway construction for the Toronto Transit Commission, dams for conservation authorities, site preparation for large industrial plants, as well as the grading, paving, and structures for many of Ontario's major highways. The intuitive style of the founding entrepreneurs has been supplemented today by proven computer-assisted planning and bidding systems, which have enabled the company to maintain its reputation as a quality contractor that completes projects on time and within budget.

The aggregates business was a natural extension for the founders, with Armbro pits and quarries supplying quality processed sand and crushed stone products to the asphalt, concrete, and building industries, in addition to gravel for road-building sites. Ready-mix concrete is supplied to construction sites in Peel, Halton, and the west end of Metro Toronto from a modern plant located in Mississauga. In 1984 United Aggregates was created with two cement companies and Armbro as equal partners. Later this operation was expanded and now markets more than six million tons of aggregates annually.

Over the years Armbro has moved into other ventures. Armbro Transport was started in 1959 with one truck and today owns and operates 75 powered units and many highway trailers. Transport and warehousing operations are managed from a terminal and office located on Dixie Road in Mississauga. Real estate developments include the 300-acre Armbro Industrial Park (originally the home of the founders' famous standard-bred horses), residential subdivisions, and rental properties. In 1975 the firm acquired Chalmers Suspensions and now markets and manufactures truck and trailer suspensions for a broad range of customers throughout North America. In 1982 Armbro entered the distribution business with its appointment as sole Ontario distributor for Deutz diesel engines, a German-designed and manufactured, air-cooled engine, used in many types of industrial and mining equipment.

Supported by a group of experienced and loyal employees, Armbro Holdings Ltd. has been managed since the late-1970s by a team of senior employee-shareholders, with the family connection continued by Charles Armstrong, Elgin's son, serving as a director.

The Lawrence Avenue Bridge in Scarborough.

H.T. WILSON INSURANCE SERVICES LTD.

H.T. Wilson Insurance Services Ltd. was established in 1927 by Harold Thomas Wilson. He has since relinquished the presidency to concentrate on farming with his son Mike, but he still retains an active interest in the insurance firm. When the company was first established, the majority of its customers were also farmers. As a result, little business was conducted during the week, but Saturdays bordered on the chaotic as all the Wilson clients would take their day in town at the same time.

The firm's original office was shared by the Ontario Provincial Police and was strictly a one-man operation. The office would later move first to Nelson, then to Kennedy, and finally to the building at 141 Queen Street East. The company's staff, which currently numbers six, handles a complete line of general insurance.

Harold Wilson was a member of the Chinguacousy Township Council for five years. He also served on the board of education for the Township of Chinguacousy and was directly involved in the original plan for what is now Bramalea.

In 1959 Joyce Bishop joined the firm; 27 years later she remains a valued employee. Joyce was on the board of directors of Grace United Church, and she serves on the board of directors of the Grace Court senior citizens' apartment.

Another valued employee of the firm is Mary McCullough. Mary has been with H.T. Wilson Insurance for 15 years handling the accounts department and is also a licensed insurance broker.

Today Austin Boland is president of H.T. Wilson Insurance Services Ltd. He became a partner in the firm in 1961. When Harold went into semiretirement in 1978, Austin took over the active running of the business. Before joining the firm, Austin was the superintendent of agents of the Fireman's Fund Insurance Company, allowing him to bring a wide range of experience to the smaller firm. He has been joined in the business by his two sons, Kevin and Danny, thus ensuring the continuity so valued by both insurers and insured.

With two boys, it is not surprising that Austin was president of the Brampton Junior "B" hockey team. He also serves as president of the Brampton Golf Club. A better-than-average golfer himself, Austin's two boys are also addicted to the game, playing out of Brampton. In 1985 Danny Boland became club champion of the Brampton Golf Club.

H.T. Wilson Insurance Services Ltd. is a relatively small firm providing a large service to longtime customers.

President Austin Boland at the birthday party with his two sons, Kevin (left) and Danny (right).

Founder H.T. Wilson cutting his birthday cake at the office on the occasion of his 80th birthday.

LANSING CANADA INC.

In 1920 F.E. Bagnall opened an office in London, England, to distribute mechanical handling equipment manufactured by the Lansing Company of Michigan. In 1930, when Lansing stopped production of industrial trucks and tractors, Bagnall secured the rights to manufacture them in the United Kingdom. Over the years the company grew, became Lansing Bagnall Limited, and by 1985 employed 550 service engineers in the United Kingdom alone. Today Lansing Ltd. is the marketing subsidiary of the privately owned Kaye Organization Limited, employing some 5,000 people.

This 1250 Steeles Avenue facility in Bramalea houses Lansing Canada Inc., a manufacturer and distributor of fork trucks, reach trucks, pallet trucks, and industrial tractors.

Lansing's Aisle Ranger, a floor-mounted crane, has the capacity to lift 4,000 pounds 60 feet high in an aisle as small as four feet six inches.

In 1954 the firm began selling lift trucks into Canada using distributors. By 1967 Lansing Bagnall had established an 8,000-square-foot warehouse and took over selling to the Ontario and Quebec markets. Warehousing was expanded to 21,000 square feet in 1969 and to 55,000 square feet five years later. The corporate name was changed in 1976 to Lansing Limited, and the Brampton group became Lansing Canada Inc. In 1979 manufacturing was begun in Brampton, and bringing the cycle full circle, Lansing Canada began to actively promote sales into the United States, capitalizing on earlier successes with highly specialized equipment including, most notably, the turret truck. The Canadian company's export business growth has been such that in 1985 it became necessary to establish a U.S. subsidiary, headquartered in Woodbridge, Virginia.

Led by its president, J. Lawrie Wharton, and backed by an innovative and diverse product mix, Lansing Canada has survived the loss of the preferential U.K. tariff rate when Great Britain joined the E.E.C., weathered the jump in value of sterling currencies and global recession, and continues to build products that are up to 75-percent Canadian. While U.S. producers have been losing market share and closing Canadian branch plants, Lansing Canada Inc. is building its U.S. sales.

The equipment produced in Brampton is electric powered, inclined to be specialized, and subject to custom manufacture. Recognizing the complementary benefit to be derived from the availability of a range of standard machines produced in large volume with high quality and low cost, Lansing Canada also arranged, in 1982, to distribute Nissan forklift trucks in Ontario and Quebec.

Typical of Lansing's innovative machines is the turret truck, which was invented in the late 1960s. Where a regular counter-balanced fork truck needs at least a 12-foot aisle and a reach truck requires from eight to nine feet, the turret truck needs only six feet. Lansing Canada Inc.'s newest model, the Aisle Ranger, is literally a floor-mounted crane, guided top and bottom in the aisle, but battery powered to change aisles. It lifts 4,000 pounds 60 feet high, is operated from an aisle of only four feet six inches, and is suitable for both pallet handling and order picking at all levels.

IVERS-LEE LIMITED

Ivers-Lee Limited is a production and packaging specialist capable of custom packaging in blister, pouch, strip, heat sealing folders, and shrink wrapping. Its customer list reads like a "Who's Who" in pharmaceuticals and cosmetics. It is a one-of-a-kind operation in Canada.

The company was started in the United States. After World War II an inventor named Volckening developed unique strip and pouch filling machines. The machines he built were so well designed and rugged that they are still in use today, and require such minimum maintenance that they will look good for another 25 years. Believing his own name would not help sell the product, he took the name Ivers-Lee out of a local telephone book.

In 1958 property was purchased on Hansen Road South in Brampton, and the firm moved its Canadian operation from Windsor into a 15,000-square-foot facility. Five years later the company expanded the building to 32,000 square feet. Even then there was some resistance to blister and pouch packaging in Canada and the United States, while the Europeans rec-

Doug Crawford, vice-president of plant operations, checking a packaging machine before going into production.

ognized the advantages of a safer, less expensive package.

In the mid-1960s a trend toward blister packaging began. Ivers-Lee Limited's first major OTC (Over The Counter) contract was for Exlax, followed by the many makers of oral contraceptives. In 1967 both the Canadian and U.S. divisions were purchased by the U.S. firm, Becton-Dickinson. Throughout the early 1970s Ivers-Lee in Canada increased its sales but continued to function with just 12 to 14 employees. The firm's only salesman in 1976 was Kenneth E. Young, current president; however, by then the staff had increased to 30. That same year Young purchased 75 percent of Ivers-Lee, and his partner, G. King Ward, purchased the remaining 25 percent.

Since the purchase in 1982 of the building next door, the facility has expanded to 52,000 square feet, business has increased considerably, and the number of employees has grown to more than 150 people, mostly from the Brampton area. The replacement value of equipment alone is in excess of $13 million.

The company is made up of a young but loyal staff. Brian Munro, executive vice-president, is 40 years old and has been with Ivers-Lee for 16 years. Doug Crawford, vice-president of plant operations, is 44 and has been with the firm for 24 years. B. Livingston, manager of technical services, has spent 21 of his 39 years with Ivers-Lee. And T. Young, vice-president of production and materials management, is only 34 years old.

Ivers-Lee employees are highly trained and motivated to ensure the highest possible standards in packaging the wide variety of products being handled. The result is a staff that takes an almost proprietary interest in a business with a family atmosphere. Each employee is included in the company's performance

Jack Rutledge, who is in charge of machine maintenance, working on a new part for one of the production machines.

One of the self-contained production areas.

bonus program, receiving a bonus in April of each year. All personnel are subject to an initial training and assessment period, followed by an ongoing performance review by their immediate supervisors.

At Ivers-Lee quality control is imperative. All incoming material is inspected and placed in quarantine. Each product to be packaged is handled in a separate enclosed room, which has been equipped with appropriate air-conditioning and air-filtration systems. The product being processed is identified on the door of each room. Immediately adjacent staging areas are provided for shipping, and the actual in-line personnel are responsible for daily cleaning of the area. The plant excludes animals, retains a licensed pest control agency, and segregates eating and smoking areas.

This care allows Ivers-Lee Limited to produce unit-of-use packaging that is tamper-evident. While no product can reasonably be called tamper-proof, it can be made so it is evident that it has been violated. Confidence is shown in Ivers-Lee by 185 customers, including Abbott Laboratories, Cosmair, Avon, Johnson & Johnson, G.D. Searle, Merrell Dow, Frank W. Horner, Ciba Geigy, Whitehall Laboratories, Procter & Gamble, Wyeth Mead Johnson, Glaxo, Chanel, Clairol, and Sterling Drug. One-third of its customers are pharmaceutical companies, one-third market proprietary drugs, and the final third are in the cosmetic business. Fully one-half of Ivers-Lee business is concerned with the promotion of either drugs to physicians or cosmetics to consumers.

In addition to Ivers-Lee Limited, Kenneth Young operates T.M.J.-Specialty Tool and Design Limited at the Hansen Road site. That firm produces the heat-sealing dies, feeding systems, hypodermic filling needles, perforators, and die cutters required by Ivers-Lee or any other packager.

Ivers-Lee Limited is a member of the Proprietary Association of Canada; the Pharmaceutical Marketing Clubs of Ontario and Quebec; the Packaging Association of Canada; the Canadian Manufacturing Association; the Canadian Federation of Independent Business; Canadian Cosmetic, Toiletry, and Fragrance Association; and Pharmaceutical Manufacturers Association of Canada.

Brian Munro, executive vice-president and general manager, checking over a production order with sales service secretaries, Laurie McVittie and Sandra Malcolm.

AIRCRAFT APPLIANCES AND EQUIPMENT LIMITED

Aircraft Appliances and Equipment Limited manufactures, repairs, overhauls, and distributes equipment for aircraft and ships from its facility at 150 East Drive in Brampton.

Aircraft Appliances and Equipment Limited (AAE) is located at 150 East Drive in Brampton. The company manufactures, repairs, overhauls, and distributes equipment for aircraft and ships. Firms with which it deals include such commercial accounts as De Havilland Aircraft of Canada Limited, Lear Siegler Inc., Strong Electric, Lealand, 3M, GE, Teledyne, and the contractors for the Departments of National Defense in Canada and the United States.

AAE, a privately owned Canadian company, was founded and incorporated under an Ontario charter in 1949 by L.V. Myslivec, the present chairman of the board. Initially its principal operation was the repair and overhaul of aircraft accessories and the distribution of aircraft electrical product accessories to Canadian owners and users of aircraft built by U.S. companies.

In 1959 the firm designed and began the manufacture of fuel filters and also pioneered the development of coalescers for the separation of water and solid contaminants from lubricating and fuel systems for marine gas turbine power plants. The design and manufacture of aircraft tachometer generators commenced in 1966, to be followed seven years later by the design and manufacture of ground power AC/DC generators and generator sets.

Aircraft Appliances and Equipment Limited is divided into three divisions, each with its own responsibilities. A fourth, the Generator Division, which catered to the industrial, agricultural, and export markets, was recently sold.

The Repair and Overhaul Division is a Department of National Defence and Ministry of Transport-approved facility for the repair and overhaul of commercial and military aircraft equipment, ground support equipment, power supplies, generator test stands, and test equipment of all kinds. The division services automatic flight controls, sensing devices, electrical power system components, pumping systems components, ground power units, motor generators, and fuel test stands.

In the field of avionics testing equipment, the company offers a Universal Avionic Component Tester, a self-contained manual tester with flexible capability, power control, signal service, and measuring devices. It also repairs, overhauls, and rewinds stators, armatures, and rotors for aircraft rotating equipment. Among the items produced for the Department of National Defence are three conversions of existing DND test stands to enable testing of the VSCF generator system on the Canadian Forces fighter-aircraft, the CF-18.

More than 26 years' experience in research and development by the Fluid Power Division has resulted in the installation of AAE equipment on many U.S. and Canadian naval ships. Installations range from simple in-line filters to duplex coalescers that automatically change over from clogged filter elements to clean ones. They come complete with service indicators, pressure and temperature gauges, heaters, automatic level controls, safety locks, and other features. The equipment can accommodate flow rates from 0.1 g.p.m. to 200 g.p.m., and many designs meet rigid military standards for shock and vibration.

Several patented designs are available in different alloys to cope with the most demanding shocks—from arctic vessels to the high-frequency vibrations of hydrofoils and surface-effect ships. Micronic filters and pipeline strainers have been produced for the most advanced navies in the world. Major filtration systems are being supplied to all U.S.-built frigates and destroyers and to Canadian frigates.

The Technical Sales and Service Division functions as a stocking distributor

Aircraft Appliances supplies the complete AC and DC electrical generating systems and fuel boost pump for the de Havilland-8 short takeoff and landing aircraft.

for products manufactured by others. Selling equipment to the Canadian aerospace industry, this division has distribution rights for aircraft electrical, avionics, and fuel accessories produced by divisions of Lear Siegler Inc. Items include AC/DC generating systems, pitch trim actuators, controllers, heading reference units, land navigation, muzzle velocity radar, radar antennae, display and transceiver units, fuel booster, and lube and scavenge pumps. In addition, a line of RFI/EMI filters are available, and, for industrial applications, the division handles self-lubricating bearings and air-conditioning systems for commercial and military aircraft.

The company is directed by a small but competent management group. W.J. White is president and general manager of AAE and also serves as president of its Technical Sales and Service Division; B.T. Dawson is executive vice-president of the firm; J. D. Young is president of the Repair and Overhaul Division; and O.E. Pavlinek is president of the Fluid Power Division.

The technicians at the Repair and Overhaul Division augment the facilities they offer by taking training from many manufacturers on the equipment produced by those manufacturers. Some of the more common units serviced by Aircraft Appliances and Equipment Limited include DC generators, AC generators, tachometer generators, motor generator sets, air-conditioning units, hydraulic pumps, rotary inverters, static inverters, converters, transformer rectifier units, voltage regulators, DC generator control panels, servo systems, galley ovens, and even searchlights.

Canadair's Challenger aircraft is equipped with a pitch-trim actuator and controller as well as a fuel boost pump system, both supplied through Aircraft Appliances.

INKAN LIMITED

InKan Limited is a company comprised of four distinct divisions offering a broad range of imaginative products and services using architectural metal and custom-fabricated tempered glass.

The InKan Architectural Metals Division manufactures custom door hardware, balustrades, fascia, wall panels, custom-designed security desks, architectural signs, and unique metal sculptures.

The RDM Glass Systems Division provides customized glass doors and entranceways, suspended glass walls, glass kiosks, and glass backwalls for squash and racquetball courts.

The InKan Contracts Division will supply and co-ordinate everything, from engineering to final installation, for major construction projects or renovations such as shopping malls and office buildings.

The Ultralite Tempered Glass Divi-

This curving brass handrail graces the lobby of the Sheraton Hotel in Hamilton.

The clock tower atop Prince Albert City Hall has been reproduced in miniature and stands in front of InKan's offices in Brampton.

sion is a supplier of tempered and heat-strengthened glass, either clear or colored, in thicknesses from 3 to 19 millimetres. Glass can be provided in straight panels or custom fabricated to the user's specifications.

InKan Limited was incorporated in April 1976 and began with a small plant on Dixie Road in Mississauga. Initially the firm's president and owner, Hugh Duke, brought only his 20 years' experience to the business, running it as a one-man operation. His first assignment was in the Calgary, Alberta Airport, and the success of this project led to additional contracts. Over the first year the staff grew to seven, and projects at the airport lasted two years. InKan Limited's first major assignment was Terminal One at Toronto International Airport.

By 1980 the company had grown to more than 35 employees, servicing customers across Canada, in England, the United States, and the Bahamas.

In 1978 Robert McNeill formed RDM Glass, and two years later it was merged with InKan Limited to become the RDM Glass Division. McNeill assumed the vice-presidency of the expanded operation.

In May 1981, recognizing the need for expanded facilities and markets, InKan opened a new 50,000-square-foot showcase office and manufacturing plant at its current location at 14 Indell Lane in Brampton. The metal-fabricating plant was modernized and expanded, and a new state-of-the-art tempered-glass manufacturing line was installed. This provided in-house supply of glass products for the company specialties such as custom-manufactured entranceways and handrails with glass balustrades. In addition, excess production was available to a growing number of outside customers.

By 1986 InKan Limited's staff had grown to 90 employees, most of whom live in the Brampton/Mississauga area and were trained by the firm. More and more, custom metal and glass products are being exported to the United States, Europe, the Middle East, and the Caribbean, generating several million dollars in sales for the company and making exports an important part of the overall operation.

The company recognizes the importance of its people and their contributions to past and future growth. A large percentage of the staff are long-term employees, drawn from a veritable United Nations of countries with a variety of interests and experiences. Plant Machine Shop foreman Jim Gourley started with InKan as a trainee in 1978, when he was only 18 years old.

While InKan Limited boasts a computerized glass cutter, the skillful hand cutting of heavy glass, 12 millimetre to 19 millimetre, is so specialized that cutter Roberto Soler was recruited from Argentina where his skills had gained international recognition.

Salesman Reg Wheeler is over 70, but from 1957 until he retired undefeated nine years later, he was Canada's Senior Men's Water Ski-Jumping Champion. Purchasing manager T.G. "Geoff" Linnell, at age 22, was troop commander of a Royal Marine Commando Unit on the beaches of Normandy on D-Day 1944. He later was asked by General Moulton to write the privately published history of the unit.

InKan Limited also takes justifiable pride in the proliferation of prestigious projects in major cities of North America. Such a job was the Prince Albert City Hall, a replica of which stands before the InKan building in Brampton. Only InKan would guarantee the original lit-from-within tempered-glass structure against extreme snow loads and wind velocity stresses. The

This metal security desk was fabricated by InKan for the Allstate Building in Markham.

Glass stabilizers provide additional support against wind deflection without detracting from this aesthetically beautiful glass entry wall at 8484 Georgia Street in Washington, D.C.

Stock Exchange building, Simpson Tower, and the Simpsons-Bay lobby, all in Toronto, used InKan glass walls.

An impressive metal security desk graces the lobby of the Allstate Building in Markham, Ontario. In the Polysar executive office in Sarnia stands a stainless-steel monolithic sculpture. The Montgomery Building in Denver, Colorado, contains a completely glassed-in executive boardroom. The Sheraton Hotel in Hamilton, Ontario, displays an InKan curving brass handrail with tempered-glass balustrade.

The building at 8484 Georgia Street, Washington, D.C., exhibits the InKan suspended glass assembly wall, acting as one unit hanging from the head of the structure, with lateral support against wind loads provided by glass stabilizers or fins. Another innovative design is the swinging door assembly entrance to the National Press Building.

InKan is currently involved in the construction of the new Canadian Chancery on Pennsylvania Avenue.

STUBBS & MASSUE LITHOGRAPHERS LIMITED

Stubbs & Massue Lithographers Limited prints and produces cartons for packaging a variety of products from food to wrappings to games. Just about everyone in Brampton has seen or used a carton printed by Stubbs & Massue. The firm makes food boxes for General Mills of Canada, cookie and biscuit packages for Christies, metal-edged dispensing boxes for Alcan and Union Carbide, pouring boxes for Dow Chemical, long-wearing game boxes, and short-lived cake boxes with waxed insides and windows. All the printing, die cutting, window covers, metal edges, and waxing are done in-house.

Stubbs & Massue was incorporated on May 5, 1971. Originally it was a partnership between Louis Massue, president, and Jim Stubbs. A year later Massue bought out the shares owned by his partner, but Stubbs continued to work with the company until 1985.

The original plant on Rivalda Road began with 20,000 square feet of floor space. By 1980 the facility had grown to 45,000 square feet of factory space and 3,000 square feet of office space. In December 1979 some equipment was moved to the present location at 240 Summerlea Road, with the main move occurring in February 1980. It was not until April of that year that all the equipment was relocated, installed, and functioning.

Typical of the problems encountered was the installation of the new Miehle six-color printing press. It was moved and partially installed when the new cement floor began to sink. The machine had to be removed. Pits had to be dug for concrete runners and the floor relaid before the machine could be properly installed.

The new plant has 90,000 square feet of floor space of which 7,000 square feet is office space. The company currently employs 94 people, mostly from the Brampton area. The plant manager, pressmen, and some shipping people have been with Stubbs & Massue since its inception.

Louis Massue started in the lithography business in 1949. In July 1963 he formed Monica Litho, which later became Maspak Limited and the agent for Stubbs & Massue. Lenmore was incorporated as a division of Maspak on November 28, 1975, and does the die cutting and finishing of cartons printed by Stubbs & Massue. Together the three organizations make up what customers see as simply Stubbs & Massue.

The investment in equipment is expensive and ongoing. In addition to the six-color Miehle press, Stubbs & Massue Lithographers Limited has installed a five-color Planeta press, a five-color Miller, three new finishing machines including an International Gluer, an International Triple A windowing machine, seven die-cutting machines, a metal-edging machine, and two waxing machines.

Stubbs & Massue Lithographers Limited's modern office and warehouse building on Summerlea Road has 7,000 square feet of office and 83,000 square feet of plant space.

SUNWORTHY WALLCOVERINGS

Sunworthy Wallcoverings in Brampton, a division of Borden Co. Ltd., is the largest single-site wallcovering plant in North America.

Sunworthy Wallcoverings is owned by the Borden Co. Ltd. Canada, making it part of an organization that is the world's largest producer of wallcoverings. Under the trade names Sunworthy, Birge, and Foremost, the firm manufactures a complete range of wallcovering products for the residential market.

The Sunworthy factory, at 195 Walker Drive in Brampton, is situated on a 17.3-acre site and contains an office, warehouse, and a 324,000-square-foot plant. It is the largest single-site wallcovering plant in North America, with an annual production of more than 20 million rolls and a storage capacity of more than seven million rolls. Its products include pre-pasted paper, dry-strippable paper, paper-backed vinyl, and metallics.

The modern Sunworthy era began in July 1927, when Wall Paper Manufacturers Ltd. (WPM) of Manchester, England (founded in 1898), amalgamated with Staunton's Limited, Toronto (founded in 1856); Reg. N. Boxer Co. Ltd., Toronto (1908); Colin McArthur & Co., Montreal (1883); The Watson Foster Co. Ltd., Montreal (1879); and two color companies. The result was Canadian Wallpaper Manufacturers Limited (CWM).

Sunworthy began its export business in the early 1920s; today that part of the business accounts for more than 30 percent of total sales.

CWM acquired Birge Company, Inc., of Buffalo, New York, in 1959, the General Paint Corporation of Canada Ltd. in 1966, and Sertex Corporation in 1970. In the meantime, The Reed Paper Group had gained control of WPM in England in 1965. Nine years later Reed restructured the Canadian operation to incorporate the Sunworthy trade name into the company name, forming Sunworthy Wallcoverings, a Division of Reed Decorative Products Ltd.

In 1976 the warehouse operations were moved to the present Brampton site, followed by the manufacturing plant, which was relocated next to the warehouse in 1982. The Birge plant was then moved to Brampton the following year. Finally, in May 1985 the Borden Co. Ltd. acquired ownership of the Reed wallcovering division and Crown Wallcoverings in England. Combined with the U.S.-based Columbus Coated Fabrics, producer of Walltex, it becomes the largest wallcovering operation in the world.

Quality control and innovation have played a large role in Sunworthy's success. The firm has introduced pre-pasted wallcoverings, was the first to use rotogravure printing, installed high-speed, heat-embossing machines, and was the first and only Canadian firm to install specially designed flexographic printing equipment. Today Sunworthy is a market leader in high-fashion wallcoverings that combine trend-setting designs and current color choices.

A machine routing patterns on wallcovering rollers in 1948.

LIFT RITE INC.

Fred Deifel is president of Lift Rite Inc. of Brampton. Today both Deifel and his company are thriving and prospering. However, it hasn't always been so. In 1954 he arrived in Canada from Germany without friends, relatives, or acquaintances, and he kept his entire capital, the equivalent of 91 cents in German coins, in his pocket. He spoke only German.

After a series of short-term jobs, Deifel landed a profitable position picking tobacco in the fall. Eventually he learned the language and became involved in all aspects of the fork truck industry, including sales, service, and manufacture. It was his experience in the fork truck industry that prompted him, after 20 years, to strike out on his own.

Lift Rite Inc. was started in a small 3,000-square-foot building in March 1977. It began with only four employees, making the pallet truck that continues to be the best-selling product in the line. Within a few years the firm expanded several times, eventually occupying 9,000 square feet. Despite an almost terminal shortage of operating capital, Lift Rite moved into its new facility at 216 Wilkinson Road in October 1979. The shortage of capital made buying modern equipment impossible, and thus the labor-intensive end product became less competitive on the market. What now takes four hours to make then took more than eight hours.

Eventually this reality led to an exchange of shares for access to unlimited capital. The needed modern equipment was installed, and production of the pallet truck jumped to 80 per day. A dealer in Louisville, Kentucky, purchased 50 of the units in June 1986, and became the buyer of the 100,000th Lift Rite pallet truck. For the occasion Deifel had special gold medallions struck. The medallions were presented to the staff and some of the largest Lift Rite customers.

In 1985, 80 percent of the firm's sales of $8.5 million was to customers in the United States. By 1986 the staff had grown to 54 employees, producing an expected $10 million in sales. Business was so good that while the company had scheduled a total holiday shut-down in July, more than half the employees were working to catch up with the backlog of orders.

Lift Rite believes the simplicity of its designs and the ease with which its machines can be serviced, coupled with a high standard of quality, account for increasing sales at prices above those of its competitors. Another reason may be the involvement of its president, Fred Deifel, who still travels across the continent and personally goes on service calls to determine where the problems are and what can be done to eliminate them in the future. The final reason may be the profit-sharing bonus offered to all employees, many of whom have been with Lift Rite since its inception.

The company has become noted for innovative product design. Two years ago it produced a low-clearance model of pallet truck, and, when asked by the military, produced a model with an even lower clearance. When the U.S. military required a special machine to lift a nuclear reactor into a space capsule prior to launching, Lift Rite completed the machine and it was immediately flown to the launch site by a military aircraft. The firm has designed and provided special lift equipment for installing rockets on aircraft wings, and is presently bidding on highly technically specified requirements for the U.S. Navy. Another project involves detailed equipment for use in European hospitals.

Because it manufactures its own hydraulic pumps and produces all the other components itself, custom-designed equipment is more readily available from Lift Rite than from producers who must buy many of their components. However, it is the standard product items that provide the bulk of the company's sales. The pallet truck, called the Fully Hydraulic Hand Lift Truck, boasts a 5,000-pound capacity and comes in numerous configurations. Standard features are low clearance and the Lift Rite lever, which allows the hydraulic system to be placed in neutral, gives freedom when moving the truck, and eliminates additional pumping.

The Lift Rite High Lifter is touted as the missing link between a pallet truck and an expensive fork truck; it offers a 2,500-pound capacity and is designed for those working with punch presses, printing presses, and conveyors. This product is available in either electric or manual lift models.

Lift Rite also offers two individual models of stackers with rated capacity of between 1,500 and 2,000 pounds depending on the load centre and other configurations. Attachments for the stacker include roller bed platforms, slip-on platforms, drum lifters, personnel safety platforms, booms with a swivel hook, remote control units, and a fifth wheel with a standard circular pull handle.

The product no man with a furniture moving wife should do without is the Lift Rite Easy Lift. It can lift up to 300 pounds on a 20-inch by 20-inch platform to a height of 52 inches. Lifting is accomplished through the use of a simple ratcheted winch. Once lifted, the load can be tipped to move on the two larger rear wheels or just rolled on the four wheels with a level load. This one product could be more valuable to the community than Lift Rite Inc.'s sponsorships or even its municipal taxes.

PARTNERS IN PROGRESS

RIGHT: Fred Deifel, president.

BELOW: Lift Rite Inc. manufactures its various lift trucks at this 216 Wilkinson Road facility.

217

RORER CANADA INC.

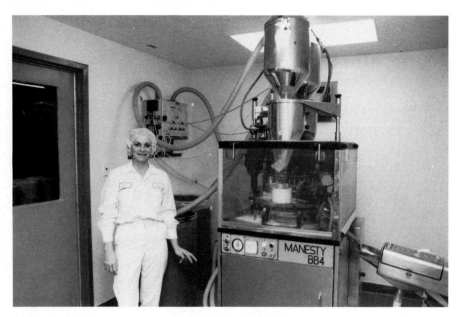

This $60,000 tablet press is used in the manufacture of Maalox tablets.

Rorer Canada Inc. is the Canadian subsidiary of Rorer Group Inc., which develops, manufactures, and markets prescription and over-the-counter pharmaceuticals worldwide. Most people will identify Rorer as the maker of the Maalox line of ethical antacid products, innovative because they neither constipate nor laxate the user.

As with so many companies in the industry, Rorer started around the turn of the century as the outgrowth of a Philadelphia-based retail pharmacy. Growth has been most dramatic recently. With the acquisition in 1986 of U.S.V. Pharmaceutical Corporation and Armour Pharmaceutical Company, worldwide sales are expected to move from the 1985 level of more than $300 million to more than $800 million in U.S. currency. Major products include gastrointestinals, cardiovasculars, blood plasma fractions, analgesics, bronchodilators, dermatalogicals, and calcium level regulators.

In discussing any pharmaceutical company it is interesting to consider the impact these businesses have on our way of life. By shortening or eliminating the need for hospitalization and providing an alternative to surgery, economic studies have estimated that Canadians save more than $700 million per year in five areas alone. Since polio vaccines were introduced in 1955, the annual savings by 1985 are estimated to be $300 million. Other annualized savings are: pneumonia vaccine, $143 million; the antiParkinsonism agent, L-Dopa, $172 million; and cimetidine, used in ulcer treatment, an additional $93 million.

The drop in the death rate for some disease categories is equally impressive. Deaths from communicable disease have declined almost 100 percent: The death rate from tuberculosis is down 99 percent; for gastritis duodenitis/enteritis/colitis, the drop is 95 percent; and deaths from influenza, bronchitis, and pneumonia are off 87 percent. To put those figures in perspective, the 16 institutions that had been devoted to care of tubercular patients had all been phased out by 1978. Had the tuberculosis rate continued even at the 1930 level, by 1982 the disease would have killed almost 20,000 Canadians per year instead of only 140.

In 1983 Canada spent approximately $33 billion on health care. Of that total, only 4.5 percent was spent on prescription drugs. The companies supplying those drugs employed more than 15,000 Canadians and pumped almost $400 million in salaries back into the economy. In addition, while development costs for new drugs have risen to approximately $100 million over 10 years and food prices have escalated 320 percent in the past 25 years, prescription drug costs have risen only 260 percent.

The history of Maalox in Canada begins in the 1950s, when Smith, Kline, and French were licensed to market Maalox in Canada. In 1968 Rorer es-

A portion of the Maalox liquid packaging line, which is capable of filling 22,000 bottles per day.

tablished itself in Canada with three employees, and by 1969 had reached one million dollars in sales. Lyle B. Goff, vice-president and managing director, joined the enterprise in 1979, coming from three years with Dow Chemicals and 23 years with Parke-Davis.

In a program working closely with physicians, Rorer Canada Inc. has built its sales to $30 million in 1986. The Brampton staff has grown to 100 in addition to 60 representatives located across the country. The company offers a complete range of educational assistance to all its employees, paying for all course books and tuition when a work-related course is completed. While the workers have been covered by a union contract for 12 years, there have been no grievances for several years.

Quality control occupies fully 17 percent of the Brampton Rorer staff. All products and packaging components must pass quality-control inspections before entering the plant. Antacid, for example, is packaged, quarantined for two weeks after production, then tested again for any micro-organisms that might have grown over this period. During processing the product is treated with a bacteriocide, but the tests are done to confirm that there is no growth.

The plant uses high-tech equipment and achieves some spectacular production numbers. Tablet production has gone from 1,000 per hour to 3,000 per hour. The blister packager turns out 150 strips of 10 tablets every minute. The machine that bottles the 12-ounce Maalox fills 22,000 bottles per day, then forwards them to a case-packing machine purchased in Orangeville.

Rorer Canada Inc. is a member of the Brampton Board of Trade, the Canadian Chamber of Commerce, the Canadian Manufacturers' Association, and Peel Personnel Association. Staff members are active in the Credit Valley Industrial Accident Prevention Association and the Brampton Industrial Association.

Lyle Goff is a past director of the Council for the Accreditation of Pharmaceutical Manufacturers Representatives. He also served as chairman of the Pharmaceutical Manufacturers' Association of Canada.

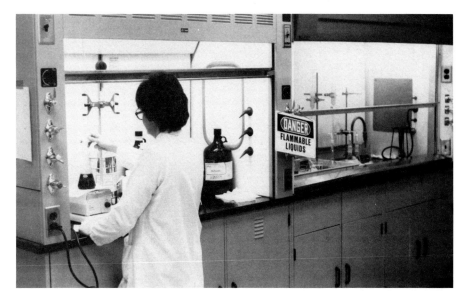
Fume hoods are utilized in laboratory testing to maintain a safe environment.

Quality assurance personnel, shown here in Rorer's Chemistry Laboratories, make up 17 percent of Brampton employees.

KNOX MARTIN KRETCH LIMITED

Knox Martin Kretch Limited is a firm of consulting engineers, planners, and landscape architects providing integrated or individual services for environmental improvement and municipal development programs. The firm was established in 1959 as a part of the Mitchell Group of Companies, led by Mitchell Engineering of London and Peterborough, England. Originally called Canadian Mitchell Associates Limited, it shared offices with Mitchell Construction Canada Limited in an old farmhouse on Dixie Road near the corner of Highway Seven, in what was then Chinguacousy Township.

In 1961 the firm moved to 70 Bramalea Road, into the same building as Bramalea Consolidated Developments Limited, then its major client. During that year an expansion plan was launched to secure projects from other parts of Ontario. Current president James Knox joined the company in 1962, and in 1969 it moved to the new Bramalea City Centre. Charles Kretch joined the firm in 1964, and Peter Martin in 1969. In 1970 Knox Martin Kretch Limited was formed and bought out the interest of Canadian Mitchell. The current principals are James A.J. Knox, Patrick J. McGrenere, Terry W. Card, and Robert D. Fleeton.

From its base in Brampton the firm undertakes projects all across Ontario—from Sault Ste. Marie to Smith Falls, from Cochrane to Cambridge, and from Pembroke to Peterborough, Peel, and Penetanguishene. Clients include the Provincial Government, Region of Peel; City of Brampton; American Motors; and development companies such as Bramalea Limited, Costain Limited, the Rice Group, Horseshoe Resort, and also the Westview Golf Club. Originally operating with a staff of four, Knox Martin Kretch currently employs more than 30 people.

Company officials give talks to local groups and clubs, and the firm and principals are members of the Metro Toronto and Brampton Boards of Trade, Consulting Engineers of Ontario, the

Knox Martin Kretch Limited, consulting engineers, planners, and landscape architects, occupies half of this building at 220 Advance Boulevard in south Brampton.

Urban Development Institute, the American Water Works Association, the Water Pollution Control Federation, the American Public Works Association, and the Ontario Association of Landscape Architects.

The firm has consistently searched for innovative and cost-effective approaches to projects whether servicing new land for housing and industry, providing municipal services to existing communities, or for municipal water supplies or waste treatment and disposal. Computer techniques are used extensively for technical and management applications.

When you consider that Knox Martin Kretch Limited has been involved in the physical development of more than 3,700 acres of land in Brampton, you begin to appreciate its contribution to the city.

DALE & MORROW INSURANCE LIMITED

Dale & Morrow Insurance Limited is a full-service general and life insurance agency. To be sure, the service can't always include having policies delivered and explained by one of the partners, but personal attention is still important. The firm handles all types of property, casualty, and life coverage for both personal and corporate clients.

It all started in the early 1920s with founder Robert F. Dale selling insurance as a sideline with milk and eggs from his farm. The earliest contract still available shows that R.F. Dale Insurance was operating in 1928.

Dale was born on a farm in Peel County and attended Brampton High School. He married in 1904; his son, Vivian, is a druggist in Brampton, and his daughter, Frances, now deceased, became a schoolteacher. It was in high school that Jack Morrow met Vivian Dale.

In 1951 Morrow joined the firm as a partner, and its name was changed to Dale & Morrow Insurance Limited. They operated without any additional staff out of the Dale house at 107 Queen Street West. When Dale finally condescended to hire a young lady to assist them, he insisted she be a farm girl—and a Liberal.

By 1961 the company had moved to two rooms above Morrison Motors at 115 Queen Street North. Jack Edwards joined the firm, and Morrow took over when Dale retired.

In the early 1970s Morrow bought out two other insurance businesses. One was purchased from the widow of Frank Richardson, a Brampton lawyer who was also licensed to handle insurance. The second was purchased from Blair Williams.

Current president Bruce Cartwright joined the firm on January 1, 1984, when Morrow retired. Cartwright had been in the insurance industry for 30 years, both as an independent broker and later an executive with Fireman's Fund Insurance Co.

The company moved from 164 Queen Street East to 24 Queen Street East in February 1985, doubling the size of its headquarters. Its staff has grown to eight employees.

Today Dale & Morrow Insurance Limited spreads its business among six major insurance companies. The firm has the pleasure of dealing with third and fourth generations of the clients first served by R.F. Dale. In order to continue with the best possible service to their valued clients, a computer system was installed in March 1986.

The success of Dale & Morrow Insurance Limited is the result of its very knowledgeable and caring employees who are committed to the service standards established by the founders.

In a typical pensive pose, Robert Frederick Dale relaxes at the Shaw and Begg fishing cabin in June 1955.

IDEAL METALS AND ALLOYS OF CANADA INC.

Ideal Metals and Alloys of Canada Inc. recently received a letter of congratulations for its contribution to the flight of the space shuttle Columbia. This letter is indicative of many of the high-tech projects in which Ideal becomes involved.

The firm is a part of the Ideal Metals Group, which sources, inventories, and provides preproduction facilities and rapid shipment of between 4,000 and 5,000 items of aluminum and 500 of copper, brass, and bronze. The product line includes various alloys in sheets, coils, tubing, plates, rounds, flats, architectural, and structural shapes, as well as an extensive range of extruded products. All are sold, with or without processing, to manufacturers in industries such as transportation, communication, building, airplanes, electrical, and aerospace, among others.

Ideal Metals began as a partnership between two men. The business was founded in 1953 by Lionel Richler and Myer Richler in Montreal. By 1962 the company was well established, and a second facility was opened in a basement in Weston, Ontario.

The Weston branch relocated three times before eventually settling at an eight-acre Brampton site in 1978. The original 40,000-square-foot facility has now been expanded to 60,000 square feet of warehouse and processing space.

In 1981 the group opened a branch in Calgary, and three years later copper, brass, and bronze were added to the previously exclusive aluminum lines. The company expanded to Vancouver with the acquisition of &+M Non-Ferrous Metals Ltd. in 1987.

The warehouse on Orenda Road features sophisticated materials-handling systems with cantilevered storage racks, overhead cranes, vacuum lifters, and sideloader trucks. Major equipment includes cut-to-length lines, precision slitters, shears, plate saws, contour saws, and an Acra-Cut precision plate saw, the first of its kind in Canada.

Ideal Metals Group employs approximately 250 people, of whom 115 are at the Brampton location. Many are long-service employees. Management is proud that, through a public stock issue in 1985, many employees were able to become shareholders and participate in the growth of the company.

By 1986 Ideal Metals and Alloys of Canada Inc. had shown a five-year sales growth of 66 percent, an increase in employment of 49 percent, and a payroll increase of 110 percent.

ABOVE: Racks of the many assorted extrusions stocked by Ideal Metals line the walls of the company's warehouse.

LEFT: Ideal Metals Group's Montreal office, circa 1950s.

BELOW: Metal-oriented professionals and specialized, state-of-the-art equipment have earned Ideal a reputation for reliability as a supplier of metals.

PARTNERS IN PROGRESS

BRAMALEA PERSONNEL INCORPORATED

Founded in 1976, Bramalea Personnel Incorporated is a personnel placement service, offering assistance in finding both permanent and temporary employees.

The firm was established in a 600-square-foot office, flanked by a real estate agency on one side and a pizza outlet on the other. Today the company occupies the entire 2,500-square-foot building. Bramalea Personnel Incorporated was started by president and owner Doreen M. Leithwood, with a small staff that included her son, Brian Leithwood, who is vice-president of sales and general manager. By 1986 the staff had grown to 20 employees, including another son, David, who is vice-president and area manager of the Mississauga office. David has considerable computer experience gained during his 14 years with Air Canada. Bramalea Personnel Incorporated also has branches in Mississauga, Meadowvale, and Etobicoke, Ontario.

The company's Permanent Division handles requests for people to fill positions ranging from the office support staff level through the professional and technical positions up to and including top management. The Temporary Staffing Division can, on short notice, provide all levels of office and industrial temporary staff. As a further service, Bramalea's Temporary Division provides needed workers for conventions and trade shows, including hostesses, demonstrators, and couponing for promotions.

Doreen M. Leithwood, founder and president.

Bramalea Personnel Incorporated provides a dependable service by taking great care in its personnel selection. Prior to referral, all candidates are interviewed and screened to meet client requirements, references are checked, and interviews arranged. A large pool of temporary help is constantly updated and on file. Bramalea even has an annual Christmas party for its temporary people who might feel left out missing the parties typically enjoyed by those in permanent positions.

The firm is a member of, actively supports, and works within both the Brampton and Mississauga boards of trade, the Brampton and Mississauga industrial associations, the Etobicoke Business Association, the Caledon Chamber of Commerce, and the Canadian Federation of Independent Business. It also supports the Association of Professional Placement Agencies and Consultants, as well as the Federation of Temporary Help Services. Staff members are actively involved in both the personnel associations of Peel and Ontario.

Doreen Leithwood was director with the Brampton Board of Trade for seven years, past-president of the Brampton Industrial Association, and vice-president of Zonta, a professional women's association. In 1986 the national board of the Federation of Temporary Help Services presented her with its Leadership Award at the annual conference. In addition, in 1985 the Zonta Club of Brampton/Caledon and the YM/YWCA gave her the Woman of Distinction (business award) for the city of Brampton.

Bramalea Personnel Incorporated's original office and present head office at 73B Bramalea Road, Brampton.

223

NORTH AMERICAN DECORATIVE PRODUCTS INC.

D. Ashton, president and majority stockholder of North American Decorative Products Inc., has been in the wallcovering industry all his working life. The company, known as Norwall, was founded in 1975. Located at 1055 Clark Boulevard in Brampton, the firm got its start with a small building and staff, and only one printing machine. There have since been five expansions, increasing plant size to more than 200,000 square feet. The company runs three shifts per 24-hour day, five or six days per week depending on product demand.

The firm has grown to be the third-largest supplier of residential wallcoverings in North America. Substrates are produced in-house by Norwall's Coating Division. With the recent addition of two surface printers, wallcoverings can be printed by flexo, gravure, surface, laminating, and the newest, a PVC expandable coater from Holland. The PVC screen machine was the first of its kind in North America, and a second is on order.

Norwall is expecting a major Canadian swing to expanded vinyl, as the PVC coating has already captured 30 percent of the European market. Where cold embossing in paper can be lost over time, the new method provides a permanent texture, is scrubbable, and, because the back is smooth rather than embossed, the wallcovering can be prepasted for the do-it-yourself market. The process requires the pattern to be printed with a combination of color and chemicals. The paper is then baked in an oven, causing a controlled chemical reaction that blows or expands, raising the desired pattern. The height of the pattern can be varied as well as the design.

The firm specializes in wallcoverings for the do-it-yourself market. Packaging can be Norwall's own or a private label, and Norwall provides the in-store merchandising aids to make its product a self-service item. Customers are provided with racks in four-foot modules, designed for wall or free-standing aisle display. Each module is self-contained with a description of the product and samples of the patterns.

Norwall products are currently being sold to mass-merchandisers such as Zellers, Kmart, Woolco, Sears, and Home Centers, as well as chains such as St. Clair Paint and Wallpaper. The firm's product is currently exported to the United States and Australia, with Japan targeted as a future market.

North American Decorative Products Inc., which specializes in wallcoverings for the do-it-yourself market, is the third-largest manufacturer and distributor in North America of residential wallcoverings. The head office is in Brampton at 1055 Clark Boulevard.

PARTNERS IN PROGRESS

REYNOLDS + REYNOLDS (CANADA) LTD.

Reynolds + Reynolds designs and sells mass-produced business forms, in competitively priced small quantities, primarily to the automotive trade. This has been the basic concept of the business since its inception in Dayton, Ohio, in 1866, and remains so with the addition of selling modern computer technology to the marketplace.

The firm was originally Gardner and Reynolds, but in 1867 Ira Reynolds bought out Gardner, and in 1939 the Grant family purchased the controlling interest they retain today. Reynolds + Reynolds' largely autonomous Canadian operation started in 1963 with the purchase of Windsor Office Supply. By 1966, 38,000 square feet were leased in Rexdale; the firm employed fewer than 50 people and needed only one rotary business forms press.

The company expanded to Belfield Road in 1968, then expanded again in 1975 to the present Brampton facilities. The plant at 2100 Steeles Avenue East originally contained 75,000 square feet of floor space. In 1982 it was enlarged to 110,000 square feet, and by 1986 the firm employed more than 300 people in 20 offices across Canada. The Brampton plant employs more than 200 people, a number of whom have been with the organization since the beginning. Demand now requires the use of seven rotary presses, and sales have increased from less than one million dollars to a projected $31.5 million in fiscal 1987 standard forms. Reynolds + Reynolds' business forms division manufactures more than 3,000 standard forms.

The press room where the business forms division prints more than 3,000 standard forms.

The computer systems division is growing rapidly. In 1960 Reynolds + Reynolds in Dayton purchased Controlomat and by 1962 was offering an electronic accounting service tailored to the auto industry. Introduction of VIM/NET, a turnkey in-house computer network system, and the firm's software development expertise encouraged Toyota and Chrysler to contract with Reynolds + Reynolds for factory communications.

The in-house computer system was upgraded in 1984 with the purchase of $20 million worth of enhanced NCR Tower computers. At the same time development was begun with General Motors that resulted in the Canadian GM Dealer Communication System. Introduced in March 1985, the system is based on the IBM PC XT and NCR hardware and provides auto dealers with exceptional hardware performance and software integration.

Constant equipment upgrading allows Reynolds + Reynolds (Canada) Ltd. to benefit the local community: Older model computer equipment, as well as obsolete paper stocks, are donated to local schools. Reynolds + Reynolds also participates in co-operative programs with local high schools that have led to the hiring of many full-time employees.

Reynolds + Reynolds (Canada) Ltd. offers an electronic accounting service tailored to the auto industry.

PEEL MUTUAL INSURANCE COMPANY

Peel Mutual Insurance Company's head office at 103 Queen Street West, Brampton.

Peel Mutual Insurance Company specializes in writing fire insurance, but since 1976 it has expanded to include auto and commercial insurance.

The County of Peel Farmer's Mutual Fire Insurance Company came into being by special statute of Upper Canada, having been presented with the signatures of more than 100 property owners who wished to place insurance with the proposed company. The petition followed a meeting in the Brampton office of a Mr. Main on April 28, 1876, at which time 15 directors were elected, representing the different townships of the County of Peel.

Originally it seems to have been intended that one company be formed to service both the counties of Peel and York, but eventually York County Farmer's Mutual was established. Unfortunately, the York concern did not last beyond a few years, and thus a number of farmers in York became policyholders of the Peel County company.

In October 1876 a Mr. Sanderson was appointed as the first agent for the firm. He was authorized to receive a commission of one dollar per application, which he had to collect from the applicant. The secretary was allowed 50 cents per policy, which constituted his salary. By 1902 agents were offered—as an incentive above their one dollar commission—an extra 25 cents if the amount of new insurance exceeded $2,500.

From the date of incorporation until 1914 the company operated from Mr. Armour's office in the Clarks Building, probably located at 13 Queen Street East. Merchant's Bank was the banker for the firm, and in 1914 Peel Mutual moved to office space above the institution, which later became the Bank of Montreal, at the southwest corner of Main and Queen streets.

In October 1955 the company's head office was moved to its current location at 103 Queen Street West. The second floor of the building was leased to the Department of Agriculture and Food until 1986, when expansion of the insurance business required that Peel Mutual take over the entire building. Both the move and required renovations were completed in the summer of 1986.

In 1976 Peel Mutual employed seven full-time and one part-time staff members. By moving into the automobile and commercial insurance fields, the organization experienced growth from 1976 to 1986 that has been exceptional. In 1983 the firm changed its name to Peel Mutual Insurance Company, which better reflects the broader insurance coverage now offered. The net premiums earned increased from $900,000 in 1978 to nearly $2.8 million in 1985. The staff increased to 15 full-time employees supported by nine directors and 25 agencies and representatives across Ontario.

LORLEA STEELS LIMITED

Lorlea Steels Limited is primarily a constructor of building walls and decks made from roll-formed metal products, generally self-manufactured in its plant at 225 Orenda Road in Brampton. The firm also manufactures and exclusively supplies a line of cold-formed structural steel channels to stock or made to a custom size and pattern.

The concept for Lorlea Steels came from Eric McLean, who began his business as a home builder in Etobicoke in 1955. Within two years a partnership was incorporated under the name Lorlea Enterprises Limited, and plans were made in the mid-1960s to enter the commercial construction field, backed by a new plant in Bramalea. Unfortunately, Eric McLean passed away before the plant was completed, but his widow and son Ronald carried out his wishes. The new, 18,000-square-foot facility housed a rolling mill and brake press equipment, producing steel decking for commercial and industrial structures.

In June 1971 the McLeans sold the company to new shareholders, and John Simonyi and the Brodeur family of Montreal became the new owners. Mr. Simonyi brought with him experience as president of Yoder Company of Canada, known internationally as a manufacturer of heavy-duty rolling mills.

At the time of the purchase Lorlea had approximately 40 employees, including the 20-man erection crew. Annual sales were one million dollars. The decision to change the name of the firm was made in 1972. Lorlea Enterprises did not seem to identify the real nature of the business, and Lorlea Steels Limited was chosen.

By 1986 the company had experienced considerable growth, with employment as high as 230 people, many from the Brampton area. The building had expanded to 41,000 square feet, housing several rolling mills and extensive auxiliary equipment required in the production of decking, siding, and structural products.

Two examples of Lorlea siding installations are the MetroToronto Center, Toronto—construction managers, Ellis Don Limited (top); and the AMC X-58 Auto Assembly Plant in Brampton—construction managers, Ellis-Don W.A. Inc. (above).

Sales are primarily in Ontario and Eastern Canada, and have grown in volume to many times that of 1971.

The range of building projects using products and construction by Lorlea Steels Limited is extensive and varied. Possibly the most impressive display of its work in Brampton is the decking on the new AMC plant. Other projects include the Darlington Nuclear Power Plant for AECL and Ontario Hydro, the Air Canada Maintenance Building in Toronto, Ford in Windsor, and, soon to come, the domed stadium in Toronto.

LEPAGE'S LIMITED

LEPAGE'S LIMITED is probably best known for the development, manufacture, and distribution of adhesive products for home and industry. In addition to its line of adhesives, the company markets LEPAGE'S sealants; REZ* wood stains and clear varnishes; Resilacrete coatings and repair products for masonry, Polycell plaster and masonry repair, and decorative products; Dow Corning's silicones; and the RUST-OLEUM* line of antirust paints and coatings. It can reasonably be argued that the LEPAGE'S mucilaginous, nontoxic, water-based paper glue comes in the Coke bottle of the glue industry. It is a rare child who can't identify the product from the dispensing bottle.

Around 1875 William E. LePage of Prince Edward Island rediscovered the art of making fish glue. Called LEPAGE'S Original Glue, the formula allowed the ready-to-use adhesive to be held in storage for months.

LEPAGE'S was founded in 1876. Between January 1880 and January 1887, more than 47 million bottles were sold. Even the prestigious Smithsonian Institution used the glue to mount specimens. Between 1880 and 1950 LEPAGE'S Liquid Glue was the world's best selling glue, and until 1941 the firm's products

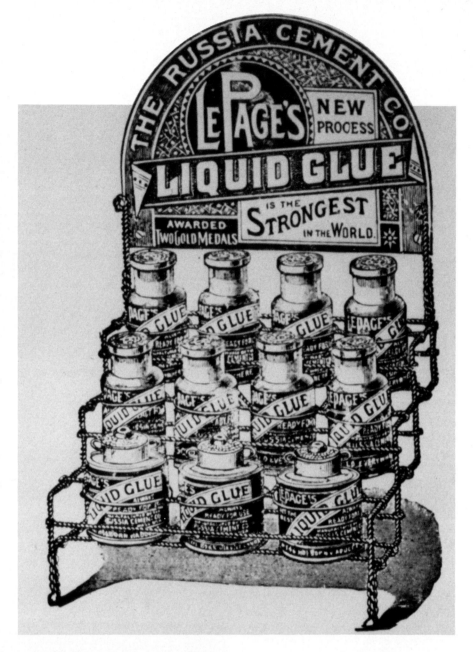

A counter rack of LEPAGE'S Glue in 1898.

A LEPAGE advertisement, circa 1876.

were sold in Canada through Gilmour Brothers & Company Ltd., Montreal sales agents.

In 1909 the search for a permanent starching agent for men's shirt collars led to the development, by Dr. L.H. Baekeland, of synthetic resin. Patented under the name "Bakelite," the electrical industry made immediate use of the product; however, its major use is in combination with wheat flour as the phenolic resin glue, set with heat and pressure, which is still the choice waterproof glue line in so-called waterproof plywood. In the 1930s I.G. Farben developed ureaformaldehyde used in plywood for interior use. That product is also sold as LEPAGE'S PANITE* Plastic Resin Glue.

LEPAGE'S Inc. was established in Montreal in 1941, and by 1951 the operation had been moved to a 25,000-square-foot office and warehouse in Toronto. Sales offices and warehouses were operated in Montreal, Edmonton, and Vancouver, with sales offices established in Saint John and Winnipeg. It was also in the 1950s that epoxy resin adhesives were introduced; representing a dramatic advancement in technology, they provide tremendous adhesion to smooth surfaces such as metal, glass, ceramics, and some plastics, while offering excellent resistance to solvents. One of the safest aircraft ever built, the Boeing 747, has more than 24,000 square feet of surfaces that are bonded with adhesives instead of rivets. The newer Boeing 767 uses even more adhesives, since aerodynamics are dramatically improved on the rivetless wings.

In 1960 the company name was changed to LEPAGE'S LIMITED, and the Canadian operation was separated from LEPAGE'S U.S.A. and LEPAGE'S U.K. In 1965 the firm acquired the RESILACRETE* masonry coating business from W.R. Spence Industries Ltd. LEPAGE'S LIMITED was acquired in 1967 by what is now known as NACAN,

"Modern" packaging of LEPAGE'S Glue in 1926.

A man cut in halves this oak stick of one square inch cross-section.

Then, with a brush, he took a drop of an amber-colored liquid and spread it on the freshly cut ends of oak, and fitted

the National Starch and Chemical Co. (Canada) Ltd., a major supplier to Canadian industry of commercial adhesives, resins, and modified specialty starches. The National Starch and Chemical Corporation is a multinational corporation with 20 plants located in the United States, Australia, Mexico, Venezuela, Holland, France, Japan, England, and Canada.

By 1971 LEPAGE'S LIMITED had acquired the REZ* wood coatings from Monsanto Canada Ltd. This product line includes wood stains (solid and semi-transparent), deck and furniture stains, wood preservative stains, and wood primer.

In 1971 LEPAGE'S moved to a new, modern 100,000-square-foot facility in Brampton. In 1977 the firm acquired the Polycell line of plaster and masonry repair and decorative products, and five years later it became a marketing agent for Dow Corning silicones. The addition in 1985 of

the stick was whole again, just as if it were still in one piece, the way it grew. He fastened one end of it to an overhead support, and on the other end he hung an automobile with two men in it. He put the whole thing on a big motor truck—

and jolted it through the streets on a long Fourth of July Parade. And the drop of amber-colored liquid held the stick together, support-

This 1926 advertisement extolled the strength of LEPAGE'S Glue.

the Canadian manufacture and marketing of RUST-OLEUM* products gives LEPAGE'S well over 100 separate product groups within six major sales areas.

Today LEPAGE'S LIMITED is operating out of a modern 250,000-square-foot plant, housing the head office, warehouse, manufacturing facility, and a research and development department that is still looking for new products, new uses, and new ways. The firm employs more than 180 people, many with 15 to 30-plus years of service.

When Brampton was advised the LEPAGE'S glue factory was coming to town, some had visions of vans arriving at the front door full of horse hooves and cow horns for the making of foul-smelling horse glue. Fortunately for Brampton, modern adhesives are mostly chemically formulated to give next to no manufacturing odor, as well as staggeringly improved adhesive qualities. Use of adhesives in aircraft is a result of tests showing that bonded structures are up to 100 times stronger than those using rivets and considerably less stressed by vibration.

Adhesives are being used in the manufacture of automobiles, bridge construction, and even in medical applications where stitches are difficult. But with all man's technology, it still hasn't been discovered how a common snail climbs a vertical teflon surface with stick and release properties more effective than anything man-made.

*Trademark

CANADIAN TIRE CORPORATION LIMITED

LEFT: Peter Montgomery, president of Peter O. Montgomery Inc., a Canadian Tire associate store.

RIGHT: Robert Bell, president of R.B. Bell (Supplies) Ltd., a Canadian Tire associate store, and past president of the Brampton Board of Trade.

Canadian Tire Corporation, Limited, was founded in Toronto on September 15, 1922, by John W. and Alfred J. Billes, who invested their combined savings of $1,800 in Hamilton Tire and Rubber Limited. Early in 1923 the firm moved first to Bloor Street, then to Yonge Street, and four years later became Canadian Tire Corporation, Limited. The first officially designated associate store opened in Hamilton in 1934, and the first catalogue was issued that same year. By 1946 there were 116 Canadian Tire stores.

By 1985 that modest original investment was producing operating revenue in excess of two billion dollars from 400 stores issuing seven million catalogues. Distribution is handled from the Edmonton distribution centre, opened in 1981; the Sheppard Avenue centre, with one million square feet of floor space; and the 65-million-cubic-foot centre in Brampton.

The first Brampton store opened in what is now Goodland's Tobacco, on Main Street North, when Les Black included CTC merchandise in his radio store. Following Black's death, his widow, Jessie, took over the franchise, eliminated radios, and moved to 20 Main Street South. Present owner Peter Montgomery, who had been managing the Davenport and Yonge Street store for then-corporate president A.J. Billes, purchased the facility in 1961. Two years later the business was moved to the defunct Handy Andy store at Queen Street and Kennedy Road, where the CTC gas bar operates today.

As business grew it was necessary to move to the firm's present location across the street at One Kennedy Road South. The 15,000-square-foot facility was expanded to 19,000 square feet in 1979. By moving the Auto Centre to 30 Kennedy Road South the following year, floor space was increased to 21,000 square feet and the Auto Centre became one of CTC's largest.

Brampton's second store opened in 1980 at 7945 Bramalea Road, just south of Steeles Avenue. The modern, 20,000-square-foot retail outlet is often referred to as the Warehouse Store, as it is right next door to one of the world's largest automatic storage and retrieval distribution centres. The store's owner, Robert Bell, has tripled sales in the first six years, and has increased its staff from the original 80 employees to 145.

Bell plays the Scottish bagpipes and collects eighteenth-century fine English antiques, some of which are gracefully displayed in his office. His civic activities include service as president of the Brampton Board of Trade, director of the Brampton Rotary Club, board member of Peel Memorial Hospital, director of the Ontario Chamber of Commerce, and marketing director of the Canadian Tire Dealers' Association. Brampton is well served by his efforts.

MANUTEC STEEL INDUSTRIES LTD.

The basic business of Manutec Steel Industries Ltd. is automotive stamping and weldments in what would be classed as a light to medium stamping mill. The firm also does a considerable amount of repair and renovation work for its neighbors in Brampton.

Manutec president Nick Lozinski is a tool and die maker by trade. As chief engineer for Lustro Steel Products he was the major customer for Beau Steel, a welding fabricating house doing about $125,000 worth of business for fiscal 1973. Lozinski purchased the company in 1974, changed its name to Manutec, and doubled sales in the first year.

The original plant consisted of 7,600 square feet of rented space at 7 Stafford Drive, with a total staff of six. In 1976 the firm moved to a 35,000-square-foot plant at 317 Orenda Road, and by 1981 it occupied 42,000 square feet at its present location, 8041 Dixie Road. A 36,000-square-foot addition was added in 1985, and by the following year the staff had grown to 68 employees. Within 12 years the building had increased by a factor of 10 and the number of employees had expanded by a factor of 11. Sales of more than eight million dollars are projected for 1986.

ABOVE: Press line up to 1,000 tons.

LEFT: Subassembly line for AMC Renault Premier.

Expansion of this magnitude has required considerable expenditures for modern high-tech equipment, such as robot welders and computers, as well as training of new and existing employees.

Over the years Manutec has built auto bumper guards, horse-trailer frames, mine drilling rigs, equipment cabinets for Bell Telephone, overhead modules for U.S. Navy helicopter carriers, a line of wood stoves under the trade name Colony Stove Works, oxygen/acetylene tank carts, and wire carriers for Sidbec-Dosco. The firm has two major customers: AMC and General Motors. On the new GM pickup truck, 12 of the highly specified parts will come from Manutec.

Surprisingly, the company does not have a single salesman in Canada. Growth has been and continues to be a result of referrals and a network of satisfied customers. However, Manutec does have a salesman in Detroit who services the U.S. auto industry.

Lozinski founded the local chapter of the Society of Manufacturing Engineers, Mississauga Chapter 212. He was the organization's first chairman, and executive meetings are still held in the Manutec offices. What started as a group of 75 now has grown to 250 members. In 1986 the society bestowed its Award of Merit on Lozinski. Manutec Steel Industries Ltd. Brampton is a member of the Brampton Board of Trade, Rotary, and the Brampton Industrial Association.

EDWARD GRAPHIC SUPPLIES LTD.

Edward Graphic Supplies Ltd. is a distributor and manufacturers' agent, providing materials and equipment to customers wishing to do flexo, dry-offset, or letter-press printing. It provides the plate making systems to take artwork from the negative stage to print.

The company was founded in 1973 by its current president, E.A. Simon. Operating out of a 2,000-square-foot plant in Montreal, the firm's three employees handled four product lines and two equipment lines, producing sales of less than one million dollars.

In 1976 Edward Graphic Supplies Ltd. expanded to Toronto and a 3,000-square-foot operation that nonetheless required further expansion by 1978; as a result, a building was leased on Gore Road in Brampton. In 1981 a new sales warehouse was opened in Vancouver, British Columbia, and three years later the head office was moved to the new 10,000-square-foot plant and office at 3 Automatic Road in Brampton. The most recent expansion was the opening of sister companies, Edward Graphics U.S. Inc., in 1985, in Seattle, Washington; Bufalo, New York; and Atlanta, Georgia.

In 1986 Edward Graphic Supplies employed 18 people, among them a number of long-term staff members. Both Kathy Simon, wife of the president, and Maida Carson in the Montreal office have been with the firm for 13 years, learning the business from the ground up. George Burns, who runs the Montreal office, has been on board for seven years, as has Joan Anderson, the office manager in Toronto. E.A. Simon's son Randy had been running the Vancouver office until he returned to Toronto to cover the southern Ontario market. Simon's daughter Laura is expected to join the business after graduating from college. Edward Graphic Supplies is an expanding company, actively seeking more good people who are interested in long-term progress with a growing entity.

Direct sales in 1986 will exceed four million dollars, and commission sales will add another $3.5 million, making Edward Graphic Supplies larger than the firm where E.A. Simon learned the business. The company represents suppliers such as W.R. Grace for photopolymers, Mosstype Corp. for mounting machines and systems, Norcross Corp. for viscosity control systems, Kidder • Stacy Co. for flexo printing presses, and Neo Industries for machine-engraved anilox rolls. It handles both products for the packaging trade and large machinery such as flexographic printing presses, selling to commercial plate makers and printers of cans, boxes, and flexible packaging converters such as C.I.L. and Peel Plastics in the Brampton area. Polyethylene, cellophane, and polypropylene are converted into wrappers for bread, milk bags, and other food packaging. Printing is done by either flexo or the gravure methods.

In 1981 E.A. Simon made his first trip to Japan and returned with plates and plate making equipment for the dry-offset

The head office of Edward Graphic Supplies Ltd. is in Brampton at 3 Automatic Road.

A product showcase printed with products supplied by Edward Graphic Supplies Ltd.

printing now used by three Continental Can plants, American Can Company, and Crown Cork and Seal, which prints all two-piece aluminum cans; and Canada Cups, which prints on foam cups, yogurt containers, and tins. Edward Graphic Supplies was the first to introduce this method into North America.

One area where Edward Graphic Supplies is pushing to change an entire industry is in the use of flexo printing in the newspaper business. By 1986 the *Toronto Star, Halifax Herald,* and the St. Catharines newspapers had all investigated the flexo system. Anyone who has had the ink from regular newspaper print rub off onto expensive suits or dresses will appreciate the absence of smears and transfers when the paper is printed with flexo equipment and water-based inks. Reported advantages for the newspaper is the lower initial capital cost; superior, sharp print results; and greater color capability. A traditional paper printed with an offset press creates 2- to 3-percent waste before the printer is certain the run is print-ready. Flexo uses water-base inks that are absorbed, giving a cleaner and clearer print. The flexo shows if it is print-ready after the first turn of the press, saving up to 3 percent of the run. On the $60 million in paper used annually by the *Toronto Star* alone, this is a dramatic savings. Edward Graphic Supplies offers both rubber and photopolymer plate systems for flexo printing.

The company has recently introduced another line. To hold polymer, natural, or synthetic rubber plates in place for printing requires a compressible, double-sided tape. Accufoam® is the registered trademark for a product designed by Edward Graphic Supplies and manufactured for the firm in Toronto that provides a superior compressible stickyback. The product is designed to reduce cylinder bounce, compensate for uneven printing plates, and results in improved print quality at higher press speeds. Accufoam® allows the presses to work with a "kiss impression," which results in longer plate life and less press maintenance. The product even offers a lower cost. Sales of Accufoam® are gaining in Canada, and shipments are being made to Europe in Spain and Switzerland, the United States, and into South America.

Edward Graphic Supplies Ltd. belongs to both the Brampton Board of Trade and the Woodbridge Board of Trade. The company's employees believe they belong in Brampton where they like the community and its people.

MACTAC CANADA LTD/LTEE

MACtac Canada Ltd/Ltee produces an extensive array of products with one common denominator: The products use a pressure-sensitive adhesive with a strippable backing sheet, or some offshoot of this technology. The firm's president, R.J. Shaw, stated that a key company strength is the ability to formulate adhesives, which are designed to perform specific functions under specific conditions.

MACtac is part of Bemis Company Inc., an international organization celebrating its 125th anniversary in 1986. MACtac was founded by Burton Morgan, a U.S. industrialist from the state of Ohio who formed the Morgan Adhesives Company in Ohio in 1959. Four years later the Canadian wing, Morgan Adhesives of Canada Limited, was incorporated in Canada, and six acres of Rice Construction's famous chicken farm on Kennedy Road were purchased. An additional seven acres was acquired later. The original 25,000-square-foot plant was constructed in 1964, and by the following year production had begun with one coating line of 15 employees.

In 1971 the company purchased a land site in Mt. Forest, Ontario, and erected a 25,000-square-foot plant at that location. Sales offices were established in Vancouver and Montreal, the name was changed to MACtac Canada Ltd/Ltee and in 1984 a plant was opened in St. Hubert, Quebec. By 1986 sales had increased to more than $40 million, and steady employment over the years was being provided for 200 office and plant personnel in a total of 230,000 square feet of plant space.

MACtac is probably best known by consumers for a product it no longer sells—the rolls of adhesive-backed decorator vinyl used to line shelves and cover walls. The active consumer products, an array of hooks and household accessories, actually now account for only 15 percent of MACtac sales. The largest portion of the business is the self-adhesive paper, film, or foil sold to flexographic, lithographic, letter-press, or screen-printers. These materials ultimately become decals, labels, price stickers, the easy-release name badges that won't stain your suit or dress, and bumper stickers.

MACtac also produces the two-sided, adhesive-coated foam tape mounts, which hold up towel dispensers, signs, notice boards, and paper cup dispensers. For the building trades, insulation foil is sold in quantities to wrap pipes and offer fire-retardant joints in pipes and ducts. For the consumer, MACtac offers household hooks, safety treads for stairs, soft flocked furniture protectors, aluminum foil tape for home insulation and auto repair, nonslip bathtub and shower treads, and the MACtac lint removing brush.

One of the most successful of MACtac's products is the Label'ope, the versatile self-adhesive pouch designed to hold such items as packing slips, courier bills of lading, promotional packages, warranties, and spare parts. MACtac Canada Ltd/Ltee produces in excess of 100 million of these products every year.

The MACtac Canada Ltd/Ltee national head office and manufacturing facility in Brampton.

PARTNERS IN PROGRESS

SHERWAY WAREHOUSING (1977) INC.

Sherway Warehousing (1977) Inc. offers a comprehensive package of public warehousing options. The company can provide storage, distribution, pick and pack order assembly, and production line product-on-demand storage, selection, and consolidation services.

The firm was started in July 1977 by founder and president Paul Rockett. Its first facility, which consisted of 18,000 square feet of rented warehouse space, was operated by Rockett with the assistance of one office clerk. In 1980 the company moved to 11 Finley Road to a facility more than double the size of its previous headquarters. That year Bill Salmon joined the firm as general manager.

The move was prompted by a long-term contract with Bristol Myers International of Belleville, a shipper of baby formulas in powder and liquid form, mainly to third world countries. Bristol Myers' goods are stored with Sherway Warehousing, then floor-loaded into containers for ocean shipment. During one seven-day work period Sherway Warehousing was able to process a rush order for Iran consisting of 95 container loads of 1,700 cases each, while still maintaining regular business.

In 1981 an additional 31,000 square feet of space was secured at 115 West

The Sherway Warehousing (1977) Inc. sign on the building in the background indicates the Finley Road location. The building in the foreground is the new facility at 1100 Steeles Avenue, which was built in 1986.

Drive. More rented space was acquired, and in 1986 Sherway purchased a new warehouse at 1100 Steeles Avenue. These additions give the company a total of more than 200,000 square feet of warehouse space in the Brampton community.

By September 1986 Sherway's staff had grown to 22 full-time employees and a pool of temporary people. Sherway utilizes eight fork trucks, and each warehouse location has its own on-site foreman. The organization specializes in assisting customers who require a high degree of special attention and service. That care is made possible by retaining employees familiar with customer requirements. Sandy McArthur has been with the company since 1980, and Barbara Rockett, wife of the president, has provided valuable assistance over the years.

Today the firm services 70 corporate clients, including Moog Canada, T.J. Lipton, Galco Foods, Molson Breweries Ontario Ltd., Catelli Foods, Dom Glas Inc., and Stafford Foods, which uses Sherway for its pick-and-pack operations.

Long-term employees Bill Gallant, Karen Lamb, Bill Salmon, Sandra McArthur, and Paul Rockett (left to right) have been instrumental in the company's ability to grow.

To date the company has never lost a customer through dissatisfaction, and credits this to its commitment to personalized service. Sherway Warehousing (1977) Inc. contends it will continue to seek and service customers requiring special attention and service levels.

235

DACHEM LIMITED

Colorwel may not be a household word, but nearly all households contain the end product of its efforts. Colorwel Limited contributes to the base materials used in children's toys, the clear dark covers on stereo equipment, the outer case for cassettes used for music and computers, pill boxes, automotive parts, egg cartons, pails, and plastic film, among others. The color, the quantity, and often the opacity of these products is all provided to manufacturers using the most modern of computer technology and skills.

The Dachem Limited and Colorwel Concentrates Limited offices and manufacturing facilities in Brampton.

President Fred Dagg (seated) with son Tom Dagg, vice-president, and daughter Jane Dagg, a sales representative with the firm, discussing future plans.

Dachem Limited converts three basic resins—polypropylene, polyethylene, and polystyrene—compounding them into useable polymers or impact sheets for vacuum forming. The firm specializes in custom color compounding and the extrusion of high-impact polystyrene sheet. Dachem also distributes a complete line of high-density polyethylene, polypropylene, and polystyrene as the Canadian agent for U.S.-based Amoco Chemicals Corporation.

Colorwel Concentrates Limited, a wholly owned subsidiary, produces color concentrates by extrusion, dry color products, and specialty concentrates that can be incorporated as stabilizers and additives. That allows other manufacturers to mechanically blend the product with their base resin to achieve both the desired color and the opacity required by their end product.

The firm's president, Fred Dagg, and several partners founded the business in 1974. The original plant consisted of 10,000 square feet of floor space on Stafford Drive in Brampton. Two years later Dagg took over the company, which is now owned by the Dagg family. His son Tom is vice-president of Dachem, and his daughter Jane is a sales representative.

In 1977 Colorwel Concentrates Limited was organized to provide color concentrates in-house. A new factory was built at 275 Glidden Road in Brampton that provides 46,500 square feet of existing floor space, with room to double that when expansion is required. The plant has a rail siding as well as the bulk-handling and storage capability to handle cars arriving from points as diverse as Sarnia and Texas.

Today the facility employs 50 people, 99 percent of whom live in Brampton. Typical is plant manager Sid Roberts, who has lived in the city for 25 years and was one of Dachem's original employees. Brampton was chosen as home for Dachem Limited because it was close to the firm's largest market and also because it offered a quiet suburban environment for employees. The plant processed a total of 26 million pounds of products in 1986, with a substantial quantity going to Kord Products Limited in Brampton for its flower pots and trays.

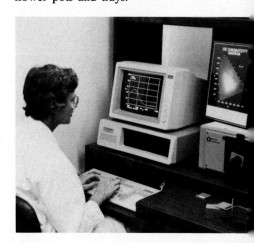

Color being computer checked to assure uniformity and match.

PARTNERS IN PROGRESS

TRESMAN STEEL INDUSTRIES LTD.

The strength of its structural steel, interlacing innumerable industrial and commercial buildings across Canada, is matched by complementary backgrounds and a depth of experience, pooled in the founding of a thriving Brampton enterprise just over a decade ago.

From design and engineering through machining and fabrication to final installation, the products of Tresman Steel Industries Limited, located at 133 East Drive, bear the imprint of the precision craftsmanship and customized service associated with the company from its inception.

Although the physical plant has been expanded by many thousands of square feet and the annual sales volume has escalated from a first-year figure of $181,000 to $11 million, the original owners remain in essentially the same hands-on roles responsible for attracting their first major contracts—the TTC subway station at Yorkdale in 1976 and General Motors' North Plant extension at Oshawa in 1987.

The president and general manager of Tresman Steel is Kasey Sarna, who brought extensive experience in all facets of the steel business to an alliance with Irme Kenedi, P.Eng., now a vice-president and manager of engineering. They incorporated with a capital of $10,000 in April 1976, and shortly afterward were joined by brothers Ivan and Wally Trentin, who were operating a miscellaneous fabrication and installation concern. Today Ivan Trentin is secretary/treasurer and plant manager, while Wally Trentin is a vice-president and installation manager.

From a starting staff of four (of whom all are still with the company), the owners have seen the work force grow to 43, comprising 4 in the office, 4 drafts persons, 2 engineers, and 33 shop employees, plus up to 50 field personnel.

Tresman Steel got its start in the Trentins' small G.I. Metalcraft building at 7171 Torbram Road in Mississauga, but sales approaching $1.5 million dictated construction of new facilities—a one-storey office and a 9,300-square-foot shop—on three acres at the present location in 1978. Since then a second level has been added to the office and more than 20,000 additional square feet have been built for production.

Sarna credits close contact with clients and quick reaction to their requirements for the firm's involvement in an impressive roster of projects such as Peel Memorial Hospital, Michelin Tire, the Canada Wonderland Theatre, and several buildings for General Motors over the years. Tresman Steel Industries Limited has also bid successfully on steel contracts from Nova Scotia to Alberta, although its principal market is southern Ontario.

Fabrication takes place in a well-equipped shop manned by highly trained personnel.

The Tresman Steel plant, at 133 East Drive, where owner/managers apply extensive expertise to design and production of structural steel.

BRAMPTON

THE MARLATT GROUP

Thorough advance planning for participation in growth areas of the economy is the operative philosophy of The Marlatt Group, a diversified Brampton company with its head office at 75 Selby Road.

From earlier successes in the roofing and siding field, the firm has broadened its horizons to include a viable real estate division and embarked on long-range plans for future corporate-owned real estate investments.

President of the group is Mary Ann Marlatt-Dennis, whose first husband, the late Irvin "Butch" Marlatt, opened a roofing business in Brampton in 1960. As the business grew, a separate office was set up in the Marlatt home to handle roofing, and later aluminum product installations in the area.

Following the death of her husband in March 1975, Mary Ann Marlatt decided to carry on the flourishing business and invited her brother, Nicholas Van Berkel,

In 1982 this facility, at 75 Selby Road, was acquired for the firm's headquarters, which services an area from Brampton to and including Metropolitan Toronto.

into partnership. They bought a bungalow on Queen Street West in 1976 as an office for Marlatt Roofing and Siding Inc., then acquired two adjacent properties and opened a small real estate office in 1979.

The roofing and siding and real estate volumes of The Marlatt Group escalated rapidly over the subsequent years, resulting in the acquisition of the former Voyager Insurance building at 75 Selby Road in 1982. An office-warehouse-showroom facility of 10,000-square-feet, it services the residential building trade in a market area extending from Brampton across Metropolitan Toronto. The Re/Star Marlatt Realty Group Inc., with 30 sales personnel directed by president Nicholas Van Berkel, shares the facility.

LEFT: Mary Ann Marlatt-Dennis, president of The Marlatt Group.

BELOW: Nicholas Van Berkel, president of The Re/Star Marlatt Realty Group Inc.

Future plans of The Marlatt Group call for franchising under federal incorporation the Re/Star trademark adopted by the realty group in 1986. All operations will soon be programmed in the group's computer system as expansion is meticulously staged in keeping with the traditional, low-key, conservative approach of management. A continuing strength will be the input of a large proportion of experienced, competent staff members.

CANBER ELECTRIC INC.

An electrical-mechanical contracting business launched in Brampton just 10 years ago has developed a high degree of capability in meeting the specialized requirements of the area's industrial-commercial sector.

Canber Electric Inc. offers a complete service encompassing consultation, design, engineering, and installation. Since incorporation on February 21, 1977, the company has completed an infinite variety of contracts for clients ranging from automated, computerized machines for manufacturers to power controls for a newspaper press. Canber—the name is a contraction of Canada and Berlin, reflecting the origins of the founders—has also responded to a host of plants and equipment relocations within a 100-mile radius of its headquarters at 271 Glidden Road in Brampton.

The formation of Canber Electric Inc. by president Manfred Stohmann, a native of Germany, and his Canadian partner Dave Osmond, who is vice-president, blended two extensive careers in the electrical field. Stohmann had apprenticed in his homeland, emigrated to Canada in 1958, and worked up to branch manager for a large Toronto electrical contractor, which also employed Osmond as field superintendent.

One of eight service vehicles operating out of Canber headquarters at 271 Glidden Road in Brampton.

Their alliance met almost instant success with a first contract in March 1977 to install automated plastic injection machines for Sheller-Globe of Canada Ltd. in its Steering Wheel Division. By early 1978 Stohmann and Osmond were attracting sufficient work to justify a self-contained subsidiary, Canber Contracting Inc., to handle engineering functions.

The total design/build service has

President Manfred Stohmann (seated) and vice-president Dave Osmond at the inception of Canber Electric Inc. in 1977.

been engaged over the years by such corporations as Dominion Glass, Gabriel of Canada, Benson & Hedges, Indal, Rockwell International, Ralston Purina, Domtar, Carlton Cards, Lustro Steel, and Metroland Publications. In 1986 Canber Electric Inc. supplied the complete design, construction, and installation of a computerized, roll-forming mill line in one of Magna International's new auto parts plants in Newmarket.

Growth of Canber Electric Inc. is illustrated by an expansion of staff to 35 from an original seven, a fleet of eight well-equipped service trucks, and a tenfold increase in gross volume over the past decade to four million dollars annually. Professional memberships are held in the Ontario Electrical League, the National Electrical Contractors' Association (Canada-United States), and the Society of Manufacturing Engineers.

J.D. BARNES LIMITED

J.D. Barnes Limited has provided professional land surveying and related services in Brampton for almost 30 years and has undertaken the surveying of more than 40,000 subdivision lots over that time. The firm has contributed greatly to the birth and development of the Bramalea "Satellite Community," working in cooperation with Bramalea Limited since the development's beginning.

The company's involvement in the development of Brampton is not limited to subdivisions, but also includes major commercial and industrial projects such as Bramalea City Centre, American Motors, the Caterpillar site on Highway 10, as well as numerous condominiums. With this background, the firm has provided many services to the general public in the form of mortgage or property surveys.

A major contribution related to the management of the city's resources has been the production of topographic maps from aerial photography (Ontario Basic Mapping program) using a totally computerized approach. These "digital" maps have become the foundation for a proposed citywide Land Related Information System that would allow the city to better manage municipal information and provide more extensive services to the public.

Since its founding in 1960 J.D. Barnes Limited has maintained a technologically innovative approach to offering its diverse land-related services. The firm is multi-disciplinary with integrated services in legal land surveys, engineering and precise control surveys, photogrammetric mapping (from aerial surveys), a computer applications service bureau, and a Research and Development Division for software development. In addition, the Land Information Services Division meets the systems design and data-processing needs of governments, utility companies, and other agencies requiring services in evaluating and implementing automated systems to manage their data and resources.

J.D. Barnes Limited is the largest survey firm in Brampton and, combined with its offices in Toronto, Mississauga, and Oshawa, has a staff of more than 200 people. In addition to its complement of highly skilled employees, the company utilizes the latest technology which has involved a multimillion-dollar investment. This technology includes the latest in digital mapping and database management systems, electronic field instrumentation, as well as in-house software development on microcomputers, which is sold throughout North America.

Under the leadership and management of Ray Visser and Carl Rooth, J.D. Barnes Limited has established itself as the premier surveying and mapping firm in Brampton.

Most of the Brampton employees are residents of the city and are active in the community as is the company. With its longtime commitment to Brampton, J.D. Barnes Limited continues to promote and provide services that will be of benefit to the continued prosperity of the city and its people.

These aerial photographs show the North Park Drive and Bramalea Road intersections and the growth in this area between 1974 and 1985. With few if any exceptions, each and every lot was surveyed by J.D. Barnes Limited.

PARTNERS IN PROGRESS

HUMBER NURSERIES LIMITED

The horticultural showplace on the eastern fringe of Brampton blooms the year round as a full-service grower and marketer responding to the diverse needs of property beautification across the metropolitan skyline.

Humber Nurseries Limited has progressed in less than four decades from a small local nursery to the largest single horticultural facility of its kind in Ontario. The flourishing business represents the fulfilment of a dream for an immigrant couple from Holland, Frans L. Peters and his wife, Sibylla, whose combined talents, ambition, and resolve in the face of hardship have nurtured the enterprise to maturity in every phase of production and sales. They have been rewarded, too, by the attraction of their sons, Frans G. and Guy, both of whom are graduates of the Ontario Agricultural College, to administrative roles in the family-owned company.

The 150,000 square feet of greenhouse under glass and poly cover and 16,500 square feet of retail-office space on Highway 50, just south of Highway 7, is a dramatic magnification of the original nursery established beside the Humber River at Mount Dennis (Weston) in the spring of 1949. There the founding couple (he was a graduate of the Dutch State Agricultural College) engaged in growing, landscaping, and retailing. In 1952 they acquired a 100-acre farm at Mono Road for the production of local nursery stock. The farm continues to supply a large annual volume of trees, evergreens, and shrubs to wholesalers, landscapers, and Humber's Garden Centre.

On two occasions the Peters family has rebounded from the ravages of natural disaster. In 1954 Hurricane Hazel levelled their Mount Dennis property. Then, in 1973, some 14 years after moving to the present 21-acre site, fire destroyed the main building and all of the equipment. Undaunted, they rebuilt each time en route to subsequent expansions of plant, product, and service.

The current structure of Humber Nurseries enfolds four divisions: Landscaping, an integral facet of the business with three full-time designers on staff; Greenhouses, producing tens of thousands of flats of annuals and perennials with computer-aided technology; Garden Centre, a 10-acre array of every requirement from tools, fertilizers, and boxes of petunias to 12-foot blue spruce; and the Nursery Operation, which includes the propagation of stock for Humber.

Functioning efficiently with this extensive resource base is a well-trained, permanent staff of 35, some with a quarter-century of service, and up to 100 employees during peak periods. Like management, they contribute to the effectiveness of the Humber Nurseries Limited commitment to its customers: "Buy from a grower . . . buy the best."

The entrance to Humber Nurseries' Garden Centre in the early 1960s, prior to major extensions of the greenhouse and retail area.

NORTHERN TELECOM CANADA LIMITED

Bramalea is the headquarters of the switching divisions of Northern Telecom Canada Limited, the country's largest manufacturer of telecommunications equipment. The fully digital switches made at Bramalea are the most technologically advanced in the world, and it is from Bramalea that the manufacturing of switches in other Northern Telecom locations around the world originated.

The company was originally established by Bell Telephone as an internal department in 1882. At that time 13 employees were assigned to the manufacture of telephones for the Bell Network. By 1895 this aspect of the business had grown so dramatically that a new company was established called Northern Electric and Manufacturing Company.

The conception of Bramalea as a "Satellite City," provided an ideal location for the establishment by Northern Telecom of a major manufacturing facility. Northern Telecom was the pioneer Bramalea industry, starting with 20 employees in a training centre opened May 20, 1961, at 317 Orenda Road. Within two months the facility had produced the first cross-bar switching equipment, and the staff had increased to over 300. This growth necessitated an expansion to 141 Clarence Street. By 1963 both locations were moved to the present site at 8200 Dixie Road.

At that time 400 employees were located on 100 acres of property on a narrow, two-lane country road with pasture lands adjacent. The starting basic plant wage was $1.38 per hour. At the time new homes across the street could be purchased for $840 down. In 1986 that same starting employee would earn $13.43 per hour; arrive on a well-maintained, four-lane, inner-city street; and expect to spend five times that $840 on legal fees alone to purchase those houses across the street.

While the Dixie Road plant was built on 100 acres, only 50 were used initially. By the end of 1963 the staff had grown to 1,000, occupying a plant designed for 1,500 employees maximum. Original projections called for 3 percent of the staff to be technically trained, and only 30 were required to be qualified engineers.

By 1986 the staff numbered approximately 4,000 people, many of whom were required to be computer literate, 20 percent were technically trained, and 400 were engineers. Of its current staff, 904 men and 215 women have been with the firm for more than 20 years, despite an accelerating rate of technological change in both product and methods. This adaptation of the Northern Telecom staff to the multiplicity and extent of change is the real story of the Bramalea plant.

Prior to moving to Brampton, Northern Telecom was producing switching equipment called step-by-step, which was a mainly mechanical method of switching telephones calls. When the Brampton plant opened, it produced the relatively new cross-bar switching equipment, which had been developed over a 15-year period in the United States, and was manufactured in Canada under licence. Over an eight-year period a more efficient technology called ESS (electronic switching) was developed in the United States and introduced at the 1967 World's Fair. Two years later Northern Telecom in Brampton introduced the fully Canadian stored-program computer-controlled system, called the SP-1, which went into service in 1971. Finally, in 1976, after only a three-year development program, the Brampton plant developed and introduced digital switches that revolutionized the telecommunications industry.

Two considerations make this evolution fascinating: first, each new system must be compatible with all existing systems, which the company must continue to service. Second, the staff at Northern Telecom has had to keep pace with rapid changes in technology. The Brampton plant has become world renowned for switching technology. By the late 1970s the firm had developed new plants in Turkey, Austria, Calgary, Texas, and North Carolina. People from all over the world come to Brampton for training in the installation and use of switching gear, and many local employees have become world travellers.

Staff up-grading is essential, and as each employee accepts new responsibilities, he or she is immediately involved in a training program spanning both classroom and on-the-job situations. While some programs take up to 30 weeks, the norm is for a constant re-vectoring in easily assimilated stages. The employees value the accumulation of increasing skills that offer better job security at Northern Telecom. The ensuing high morale leads to many father/son, mother/daughter, and even grandchild combinations among employees.

In 1986 Northern Telecom introduced the ultimate in assembly lines with its flexible manufacturing, computer-operated automated PCP (printed circuit panel) production line. Using a system of bar codes and readers, printed circuits of all varieties can be assembled automatically on the same software-driven line, and provide the quality required of equipment in this modern age.

The Bramalea plant at the steel framework stage in June 1962.

PRO PRINT BRAMPTON

The staff in front of Pro Print's Selby Road facility.

Pro Print Brampton offers a complete commercial printing operation. The company handles a wide variety of printing requirements extending from business cards to books, brochures, annual reports, banners, presentation folders, and letterheads from one to four colors. Pro Print also does extensive labels and decal work through its screen printing department. The office supplies in stock are quite extensive, with most orders being processed within two days.

Pro Print was founded in May 1980 by its president, Bill Carter. Originally the firm operated out of Carter's town house with a staff of two, but within a month had to expand. A customer provided 300 square feet on his own premises at 97 Orenda Road in Brampton, and within a year the staff had grown to seven, the space required to 2,250 square feet, and the customer base to almost 300. The customer/landlord eventually moved out to provide the space required.

On January 1, 1985, Pro Print took possession of its current facility at 25 Selby Road, expanding to 4,535 square feet, adding the office supplies business, more than doubling the customer base, and employing 16 full- and 4 part-time

Bill Carter, president.

workers. Bill Carter credits much of his firm's success to the dedication and hard work of his employees.

Today the management of Pro Print is made up of four partners. Bill Carter is president, bringing to the firm 14 years of experience in the printing business, from hands-on production to sales and marketing. Vice-president Dennis Maister joined the company in September 1984, and secretary/treasurer Paul Higgins came on board in February 1986 as well as Richard Burley in March 1987. Another key staff member is plant manager Paul De Brou, who started with Pro Print in November 1982.

Because the diversity of printing services offered by Pro Print requires it, each member of the staff must bring a special blend of flexibility and "people awareness" to the job at all times to provide the needed customer consciousness. The firm is working toward specializing in the high-tech printing jobs often shunned by others.

Pro Print can screen print jobs on the premises and offers fully staffed screen, art, press, and bindery departments to its growing list of customers, which ranges from the individual off the street to major breweries and local aerospace high-technology companies.

Plans are already under way for future expansion, providing present staff can be trained and new staff acquired to guarantee that the level and quality of service remain high. Pro Print Brampton continues to rely on word-of-mouth advertising to expand its customer base.

WELLWOOD CONTAINERS LIMITED

Wellwood Containers Limited produces corrugated containers. Servicing a specific portion of this market, it concentrates on the efficient production of regular slotted cartons, die cuts, and other styles of cartons from corrugated sheets purchased locally. Wellwood is capable of printing the boxes in any color, and is currently investigating the feasibility of stick-on multicolored labels. The intention of the company is to build the volume of straightforward cartons rather than expand into additional varieties and complexities.

The firm was founded in 1984 by Gary Wellwood and Donna Campbell. The plant, located at 31 Strathearn Avenue in Brampton, is large enough to accommodate future growth. Originally the staff consisted of only three people. However, by 1986 its staff had grown to include 12 full-time employees and eight-plus part-time workers during production peaks. Wellwood selects his staff carefully, looking for people who are personable, sensible, and will respond to the firm's training program. This is possible thanks to 50 years of on-board experience among the management staff.

Gary Wellwood, president, has 15 years' experience in the corrugated industry, including inside management as well as a solid sales background. Secretary/treasurer Donna Campbell brings over 10 years' experience as a general manager; prior to that, order desk sales; and has previously worked in the industry with Wellwood. Plant manager Ed Gill has worked in many different areas of corrugated for over 10 years providing related production experience. Don MacDonald, originally a designer, has recently joined the firm as sales representative, bringing an additional 15 years' corrugated experience to Wellwood.

Many of the Wellwood Containers Limited staff are from the Brampton area. The company offers opportunities to local students after school and during the summer months, and has also created emergency second shifts using local women on a part-time basis.

While the corrugated industry has a considerable number of large suppliers, it is the smaller suppliers, providing service and attention to smaller manufacturers, who control fully 50 percent of the market. Wellwood Containers Limited has sufficient faith in the industry and its growth to invest in its first brand-new truck of what hopefully will become a fleet in the future.

The home of Wellwood Containers Limited and the first of the company fleet.

Ed Gill, plant manager, feeding the folder gluer.

NABISCO BRANDS CANADA GROCERY DIVISION

The Nabisco Brands Canada Ltd. plant in Brampton specializes in the production of Milk-Bone dog biscuits and Flavor Snacks. With almost 80 percent of the dog biscuit market, Milk-Bone is far and away the best-selling dog biscuit in Canada. Nabisco in Brampton is one segment of the Nabisco organization, which provides us with Nabisco cereals, Cream of Wheat, Christie's and Peak Frean Biscuits, and Planter's nuts.

Nabisco began in 1898 when two groups of independent bakers formed the National Biscuit Company in Chicago. The first Shredded Wheat biscuit was made in 1901, and the Canadian Shredded Wheat Company in Niagara Falls was acquired in 1928. The company name remained the same until it was changed to Nabisco Foods Limited in 1954.

The Milk-Bone division started when Nabisco acquired the Ross Miller Biscuit Co. of Napanee, Ontario, a producer of biscuits for dogs and foxes. The plant had been closed and, under Nabisco, reopened with 10 employees in 1953. Packaging was changed to modern methods, but the biscuits were made using an old-fashioned cutting machine and coke-fired reel oven that operated like a Ferris wheel with 12 rotating shelves. Milk-Bone was first produced in 1955, becoming an instant success, and Milk-Bone Flavor Snacks were added to the line in 1965.

The need to modernize and expand led to the opening of the existing plant at 70 West Drive in Brampton in 1971. Of the Napanee employees, only the plant manager, Fred Keith, made the move to Brampton to head a staff of 38.

The biscuits are imprinted using brass roller dies, then baked on a wide band oven measuring 125 feet long and 4 feet wide, with temperatures ranging from 400 degrees to 600 degrees Fahrenheit. Whereas the Napanee plant produced 481,000 biscuits per day, the Brampton plant produces 1.5 million per eight-hour shift. By 1986 the plant had added equipment for handling bulk storage, such as the flour drawn from the Streetsville plant. The facility employs 38 people, most of whom live in the Brampton area.

In addition to possessing a taste that dogs prefer and doing a recognized job of cleaning the teeth, the Milk-Bone biscuit is also a 100-percent nutritionally complete dog food. The company is so certain of this fact that it guarantees in writing on every box that its Milk-Bone biscuits meet or exceed the nutritional requirements of every stage of a dog's life cycle. The product includes everything from wheat germ to vitamins.

Nabisco is proud of its Milk-Bone Warriors hockey team; it has championed the industrial league for four of the past five years. The firm also actively supports and sponsors a minor baseball team in Brampton, and belongs to the Rotary Club of Brampton and the Brampton Board of Trade.

Ontario Premier William Davis cutting the ribbon of Milk-Bone biscuits to officially open the Brampton Nabisco plant on May 12, 1972. With him are Ike Williamson (right) and John Whitton.

The Nabisco Brands Ltd. Foods Division facility at 70 West Drive in Brampton.

PARTNERS IN PROGRESS

VOYAGEUR TRAVEL INSURANCE LIMITED

Voyageur Travel Insurance Limited is a Canadian agent for Lloyd's of London and specializes in travel insurance. The firm is a leader in the field, providing five main types of travel insurance. Trip cancellation covers the traveller for the non-refundable portion of air fares and package tour costs in the event of sickness or death. Out-of-country medical expenses covers those costs over and above what is covered by Provincial medical plans. Air flight accident insurance, common carrier and vehicle accident insurance, and baggage are the other types of travel insurance that are offered by Voyageur. The firm's business is handled from the Brampton head office and its branches in Vancouver, Calgary, Winnipeg, Montreal, and Halifax.

The company began as a concept of founder George A. Underbakke in mid-1966 in the basement of his home at 19 McClure Avenue in Brampton, with two employees and a printing press for printing forms, brochures, and cheques. In 1967 the company was incorporated, and since then steady year-on-year growth has been maintained. By 1973 Voyageur had outgrown the basement, and the firm bought a new building at 75 Selby Road. Branch offices were established all across Canada through the 1970s, and in 1980 the company moved to the current headquarters at 44 Peel Centre Drive.

Toward the end of 1980 Crawley Warren (Holdings) Limited of London, England, became majority shareholders of Voyageur Travel Insurance Limited. Today the firm has a staff of over 150, 75 of whom operate from the Brampton office. A large sales force is employed along with administrative, data processing, and claims personnel, who handle all aspects of the business except for financially underwriting the product line, which is done at Lloyd's. When the company opened in 1966, the largest portion of its business was in baggage and accident, but with the growth of non-refundable and limited validity air fares, coupled with the staggering cost of hospitalization in the United States, the needs of travellers have changed and the firm has responded.

Sales are handled through a network of more than 5,000 travel agent offices, including many of the major players in the industry. Voyageur's first catastrophe loss resulted from the downing of Korean Airlines flight 007, and other losses were sustained as a result of the Air India disaster over the Atlantic.

Voyageur Travel Insurance Limited got its start in Brampton and expects to maintain its head office location in this city. The firm is a member of the Brampton Board of Trade and ACTA (the Alliance of Canadian Travel Associations). Its executive vice-president, John W. Hudson, has served for 12 years on the Peel Memorial Hospital Board, two of those years as chairman.

The head office of Voyageur Travel Insurance Limited is located at 44 Peel Centre Drive. Senior management (left to right) are Peter D. Fairgrieve, vice-president/finance; John W. Hudson, executive vice-president/chief executive officer; and John D. McKenna, senior vice-president/sales and marketing.

McCLEAVE TRUCK SALES LIMITED

In 1987 McCleave Truck Sales Limited celebrated its 50th year as the International Harvester dealer in Brampton. Shown here in 1939 (from left to right) are Jack Calvert, Mr. Hilliard from International Harvester, Jim Calvert, Mr. Allan from International Harvester, Walter E. Calvert, and "Mac" McCleave, founder of the dealership.

McCleave Truck Sales Limited is Canada's largest Navistar (formerly International Harvester) dealership, and in 1987 celebrated 50 years of continuous association as an International truck outlet.

The firm was founded by J.A. "Mac" McCleave. McCleave served his apprenticeship in Ottawa in 1913, then moved to Brampton in 1920, by which time he was a master mechanic. His first garage was started behind his home on Mill Street South, and his first dealership was with Durant automobiles. He later took on a Hudson dealership; but, when the chance was offered to become the local International Harvester dealer in 1937, McCleave began the relationship that has lasted to the present.

The dealership was moved to 250 Queen Street West in 1940. When Mac McCleave sold the business in 1958 to 20-year-old Al McCleave, its current president, the company employed four people. By 1965 the business had expanded to its current head office location at 36 Rutherford Road South. Three years later the firm diversified into truck leasing, which now accounts for more than 30 percent of the business. The leasing operation has its own distinct corporate offices at 110 Rutherford Road South.

In 1979 Al McCleave opened the dealership's first branch in Orangeville, Ontario, and began the expansion of the Rutherford Road location, doubling the size of the facility by 1980. That same year saw the opening of the McCleave truck body shop at 19 Davidson Road in Brampton. Succeeding years brought additional expansion. First a branch was opened at 3075 Lenworth Drive, Mississauga, in 1983, then a Concord branch was opened the following year. When Chris McCleave became vice-president of the Parts Division and corporate legal counsel in 1984, he represented the third generation of his family to be involved in the business. In 1986 the company added leasing facilities in Montreal, Scarborough, Concord, and Mississauga.

McCleave has a staff exceeding 200 people, mostly from the Brampton area. The firm has in excess of 600 leasing trucks in its fleet, serving many corporate accounts including Martin Brower, who distributes food from Winnipeg to Ottawa, wherever the golden arches of McDonalds takes it.

As a master mechanic by trade, Mac McCleave placed a high priority on service to customers of the dealership. This practice continues to be upheld by both the sales and the leasing operations of McCleave Truck Sales Limited. Mac McCleave also contributed to Brampton by serving on council through the late 1940s and as a founding member and the first president of the Brampton Flying Club.

Ted Armstrong, Mr. Hilliard, Mr. Allan, and Mac McCleave (from left to right), and Huck Ridler (in driver's seat of truck), pause during their business transactions to pose for this July 1939 photograph.

PARTNERS IN PROGRESS

TEXPORT, A DIVISION OF TNT CANADA INC.

Texport is a division of TNT Canada Inc., which is, in turn, a wholly owned subsidiary of TNT Ltd. with head offices in Sydney, Australia. TNT is the largest land transportation firm in Canada, comprised of 12 separate companies including Texport, Overland, Kwik-As-Air, All Trans, and Champlain, among others. Texport offers a complete transportation and distribution service, exclusively to the fashion/garment industry, and is the only national distribution company in Canada to offer the garment-on-hanger distribution method.

Texport began in October 1974 as a concept to bring the hanger-garment system used in Europe to Canada, where the tradition was to ship garments boxed. General manager Paul Nelson and three employees started with the firm, which was backed by Abitibi/Price and Lever Brothers, whose Unilever associate company shipped garments on hangers in Europe. The first plant was a 3,000-square-foot facility on Viscount Drive in Mississauga. By 1975 Texport had moved to a 10,000-square-foot facility on Northwest Drive, sharing the space with Abitibi/Price. Within two years the company relocated again, this time to a 20,000-square-foot building on Kimbell Street.

In 1978 TNT purchased Texport, which had sales of four million dollars. Since then dramatic growth has led to expansion into a new 70,000-square-foot facility in Brampton, with additional space available to double its current size. The move to 230 Advance Road took place in 1985. This facility houses Texport's head office and 130 of its 300 Canadian employees. The transportation and distribution company has other facilities in Montreal, Winnipeg, Calgary, Edmonton, and Vancouver; runs a fleet of 200 trucks; and has agents in Maritime Canada, Europe, and the Orient. Sales in 1986 exceeded $25 million, distributing in ex-

Brampton Mayor Ken Whillans (second from left) cuts the ribbon at the 1985 opening of Texport's new Advance Road facility. Celebrating the occasion are (from left) G. Rowe, P. Nelson, and J. Smith.

cess of three million garments internationally to complement Canadian volumes.

Texport offers TEXPRESS, an express road service to and from Western Canada; Texport Air Forwarding and Sea Forwarding from fashion centres such as Milan, Paris, Rome, and London; as well as a broad range of pick-and-pack, storage, inventory management, consolidation, bonding and sufferance areas, customs brokerage, and even the stuffing or destuffing of air and sea containers.

Forty percent of Texport customers are smaller independents, with some of its larger customers including Sears, Marks and Spencer, Eatons, Liptons, Penningtons, and Simpsons. Movement of all goods is supervised by a highly computerized system capable of providing up-to-date tracing both quickly and accurately.

Texport believes in supporting the fashion industry, offering a fashion information and news bulletin called "Texport Folio," and actively promoting The Canadian Festival of Fashion in Toronto.

From this 70,000-square-foot facility, Texport delivers garments on hangers nationwide, the only garment distribution company in Canada featuring this service.

HAESSLER-DEWAY LTD.

Haessler-DeWay Ltd. began as a small machine shop serving Brampton area industrial customers, and quickly evolved into the design and manufacture of special-purpose machinery and equipment. The company supplies packaging and food-processing machinery, special material-handling systems, a variety of special-purpose machinery, and the Skyjack line of aerial platforms.

Wolf Haessler, German born and educated in Canada as a professional engineer, and Sid DeWay, an experienced welder and machinist from Holland, founded Haessler-DeWay Ltd. in March 1969. The business started with 2,000 square feet of rented space at 67 McMurchy Avenue South in an old tannery building behind the Bi-Way discount store at the same address. (By sheer coincidence, the tannery building burned down accidentally in July 1986, on the day that Haessler-DeWay Ltd. held an open house celebrating its fourth relocation to its present facility at 45 McMurchy Avenue North.) By 1970 the firm had expanded to 4,000 square feet at the original location with a staff of five.

In 1972 the company moved to a more modern free-standing 5,000-square-foot industrial unit at 75 Stafford Drive. While at this location, Haessler-DeWay Ltd. supplied, in addition to some other major production equipment, an extensive conveying system to the expanding of Kitchens of Sara Lee (Canada) Ltd. in Bramalea. The number of employees at the Stafford Drive location grew from 8 to 18.

Continued growth forced the firm to move, in 1976, to 22,000 square feet of industrial space at 100 Rutherford Road South, where its staff increased to about 25 machine designers, machinists, welders, machine fitters, and office workers. Major projects designed, built, and shipped from

This photo of the Skyjack (model SJ-1007), with a six-foot crank-out platform extension, shows the height and range of the machine.

The original tannery, the first home of Haessler-DeWay Ltd.

this facility were automatic bakery tray proofers and complex automatic packaging lines sold to customers in Canada and the United States. The 1981 to 1983 recession was survived with the help of a large and timely order for 32 cooling conveyor systems supplied to the government of Egypt for the large-scale automatic production of Pita bread. This project, having been financed by the United States, provided several opportunities for management to visit Washington, D.C., and learn about big government with regard to its management and financing of overseas projects. (In spite of many frustrations and delays this was a successful project both in its original purpose of providing a low-cost food supply to the people of Egypt and as a financial success for Haessler-DeWay.)

The baking industry was undergoing major changes in the early 1980s with the trend of consumers' preferences for baked goods going more and more toward in-store-baked specialty products, causing volumes for large centrally located bakeries producing traditional products to decline. This meant, to a large degree, a redirection of the talents and resources of Haessler-DeWay Ltd. toward new products that would ensure a future for the company and its employees.

Fortunately the need for product diversity had long been recognized by Wolf Haessler and Sid DeWay. In 1984, when it was felt that there existed a growing need for industrial aerial lift equipment, the groundwork for the manufacture of such products, laid as early as 1981, was instrumental toward what soon turned out to be an extremely successful product line of elevating scissor workplatforms called "Skyjack." The founders' foresight is evident from the fact that the wholly owned company, Skyjack Inc., was already incorporated as a separate entity in the fall of 1981, although production was not to begin until three years later.

The Skyjack elevating workplatforms are self-propelled, scissor-type, hydraulically actuated, elevating platforms used primarily by mechanical building trades and contractors for the installation and maintenance of overhead services and equipment in large buildings, factories, warehouses, and arenas. The units are manufactured with a variety of optional features, capacities, and sizes.

In order to make best use of the company's capabilities in machinery design and efficient production, it was decided to contract exclusive marketing and distribution rights of the Skyjack product line to a company that had an established record in the sale of related aerial lift equipment. This requirement was ideally met by L.W. Matthews Equipment Ltd. of Weston, Ontario, a well-respected equipment sales and rental firm that had introduced scissor-type aerial lift equipment into Canada from the United States in the mid-1960s. At this time the two original owners agreed between themselves that their personal goals and ambitions would best be met by Sid DeWay selling his ownership to Wolf Haessler. Today Haessler-DeWay/Skyjack ownership is shared 95 percent to 5 percent between Wolf Haessler and his brother Hart. The founders of Haessler-DeWay Ltd. take pride in their valued 20-year relationship that has helped to bring the company to where it is today. Sid DeWay is continuing to contribute his talents by managing the firm's computerized machines operations.

In February 1986 the mayor of Brampton, Ken Whillans, among other distinguished guests, helped Haessler-DeWay Ltd. and its employees celebrate its 100th Skyjack. Three months later the company moved to its present location at 45 McMurchy Avenue North, where it has 35,000 square feet of office and manufacturing space. As of April 1, 1987, the facility is owned by a group of 20 key Haessler-DeWay employees who lease it back to their employer, an original and unique employee/employer concept conceived and organized by Haessler-DeWay's people-oriented production manager, Giovanni Tucceri.

It is also fitting that the company and

Self-propelled elevated Skyjack workplatforms dramatically lined up at the Canadian plant of Haessler-DeWay Ltd.

its employees will celebrate, with their customers, the 500th machine within a few weeks of this book going to press. Although the phenomenal growth of the Skyjack product line today predominates in Haessler-DeWay's production, the firm continues to manufacture special-purpose machinery for a variety of industries.

The corporation, whose first machinery sale was a fertilizer spreader converted for the city of Brampton to winter salt spreading, now has 65 employees, an annual capacity for the production of over 500 Skyjack workplatforms, and orders in hand for projects with Ontario Hydro, General Motors, Chrysler Corporation, and American Motors. Haessler-DeWay Ltd.'s future promises to be bright and exciting, not so much because of its orders on hand and the quality of its products manufactured today, but because the personal dreams and ambitions of its employees are in unison with their company's purpose to produce the quality products of tomorrow.

CATERPILLAR OF CANADA LTD.

Caterpillar of Canada Ltd. was originally established to provide greatly improved parts service to the distribution network. Today the Brampton plant produces Caterpillar's worldwide supply of three models of log skidders; is the supply source for the western hemisphere of small wheel loaders, integrated tool carriers equipped with interchangeable tools and attachments; and also supplies the basic machine platforms for tree harvesters. Eighty percent of production is destined for export, mainly to the United States, but also to Latin America, Africa, Europe, and the Far East, producing significant foreign exchange revenues for Canada. Brampton is the Canadian distribution centre for a new series of Caterpillar lift trucks.

Caterpillar of Canada was formed in 1956, beginning operations in a 48,000-square-foot parts warehouse the following year. In 1963 a 63,000-square-foot manufacturing plant was added for the production of wheel loaders and motor graders for the Canadian market. By 1967 another 64,000 square feet of floor space were added to the parts warehouse, and production had begun on lift trucks.

In 1975 and 1976 manufacturing and warehouse space was increased to a total of 320,000 square feet (7.3428 acres) servicing only the Canadian market. The Mississauga plant employs 220 people and converts plate steel, purchased in Canada, into welded and machined wheel loader and log skidder components, currently being shipped to the Brampton plant.

In 1980, 200 acres were purchased in Brampton, and in three years the plant was completed, providing 480,000 square feet of office and manufacturing space. The facility employs 250 people, including many new Canadians, and, between Mississauga and Brampton, is estimated to create jobs for another 1,000 people at companies supplying Caterpillar with goods and services.

The export of more than 1,500 machines from the Brampton plant in 1984 made Caterpillar the largest exporter of construction equipment in the country. With the addition of new products, exports in 1987 are expected to exceed 1,800 units.

The Caterpillar of Canada Ltd. plant utilizes modern production technology with automated machining complexes, unmanned paint stations, computer-controlled storage retrieval systems, and computerized final test inspection of the finished machines. T.V. Ontario used the plant for an in-school promotion on the future of computer programming in industry.

The largest exporter of construction equipment in Canada, Caterpillar of Canada Ltd. occupied this Brampton facility in 1983.

The examples of Brampton-built Caterpillar construction equipment shown here are the model 508 log skidder (left) and the model 926 loader (right).

HARE REAL ESTATE (PEEL) LTD.

Hare Real Estate (Peel) Ltd. has come a long way since its inception in 1950. In August of that year Harold S. Hare founded the brokerage firm in his residence at 34 Harold Street. By 1952 the company had grown to three employees and had established an office at 82 Main Street North. In June 1964, having outgrown its Main Street location, the firm moved to new offices at 134 Queen Street East. Ten years later a group of Brampton businessmen that included Fred H. Ruff, president of Hare Real Estate; Louis Acri; Don Seeback; Randy Longfield; and Holmes Matheson demolished the original structure and put up a new four-storey building at 134 Queen Street East. All are equal partners in the new building, and all are also tenants.

In 1965 the business was purchased by Bert C. Reynolds of Brampton. Reynolds in turn sold the company in June 1970, at which time its name was registered as Hare Real Estate (Peel) Ltd. In 1972, while he was president and general manager of Hare Real Estate, Albert C. Waldrop also served as president of the Brampton Real Estate Board of which Hare had been a founding director. In fact, many of the original board meetings were held in Hare's recreation room.

Fred H. Ruff, who has been with Hare since he entered the real estate business in 1966, is the current president and owner of Hare Real Estate (Peel) Ltd., having purchased it in 1978. Ruff served as president of the Brampton Real Estate Board in 1977 and again in 1985. He also is a charter member of the Brampton Kiwanis Club and an active member of the Brampton Board of Trade.

The original staff of three in 1956 grew to 10 in the 1970s. Today the firm employs 21. The company's services have also expanded and now encompass a complete real estate service, including residential sales, commercial and industrial sales and leasing, appraisals, property management, and mortgages. Fifteen percent of Hare's business is in the commercial and industrial leasing division, with the balance weighed heavily toward residential sales.

Hare Real Estate (Peel) Ltd. is a member of the Brampton Real Estate Board, the Ontario Real Estate Association, and the Canadian Real Estate Association. The firm is also the exclusive Brampton member of A.P.R.S. (All Points Relocation Service), which claims to be able to move people practically anywhere on the North American continent with a minimum of inconvenience to the client.

The Centre and Queen Building was constructed in 1975 and Hare Real Estate (Peel) Ltd. occupied its half of the main level at that time.

JOHN LOGAN CHEVROLET OLDSMOBILE INC.

When what became the John Logan Chevrolet Oldsmobile franchise got its start in 1925, it was the only General Motors franchise in Brampton. Today it is the largest GM dealership in the area and probably the largest for any manufacturer.

Initially a Chevrolet franchise was granted to J.T. Farr & Sons, operating out of the original building located on Main Street North. Annual volume was 90 new cars and trucks. In 1928 an Oldsmobile franchise was granted, and two years later the Pontiac Buick line was added. In a photograph of Farr's Garage, circa 1928, the sign reads "British American Gasolene," reflecting the old-fashioned complete sales and service combination we see coming back today.

By 1952 General Motors had changed its dealer policy, and the Pontiac Buick portion of the franchise was transferred to Morrison Motors, also of Main Street. When the business was sold to Howard Elliott in 1962, the address remained the same but the name was changed to Howard Elliott Motors Ltd. Three years later Elliott moved the franchise to 241 Queen Street East and built a modern, new facility that included a 7-car showroom, 14 service stalls, and 10 body shop stalls.

Failing health caused Elliott to sell the franchise in 1972 to John H. Logan, current owner and president, and the name became the now-familiar John Logan Chevrolet Oldsmobile Inc. The next year saw the annual volume rise to 2,000 new and used cars, with 68 people on staff. During 1974 an expansion program was begun with 700 square feet being added to the used car building. That was followed two years later by the addition of 8,500 square feet to the service area, 3,500 square feet to the parts department, 500 square feet to the leasing office, and a 35- by 15-foot car wash. Cars serviced by John Logan Chevrolet Oldsmobile usually are returned to their owners washed at no extra charge.

Currently there are 106 people employed by Logan, six of whom have been with the company for more than 20 years. Mary Jo Whaley, secretary/treasurer, has been with the firm for more than 38 years. A million-dollar expansion was undertaken in 1985 to house the dealership's growing number of employees. The indoor service reception area added 2,500 square feet, in addition to 2,500 square feet of second-floor offices and 4,000 square feet for the showroom and used car sales offices. The franchise now covers more than 5.5 acres, and the recent purchase of adjacent property will add almost another acre. Total dollar sales from sales, service, parts, and leasing will exceed $35 million in 1986.

This enviable record of expansion has garnered numerous awards for the dealership. The franchise has won every annual GM Sales Incentive Award since 1975. In addition, it has won the GM President's Triple Crown Challenge four times, and has been the recipient of both the Select Service Award and the Truck Leadership Award. In 1985 John Logan was selected by the Toronto-area dealers as the winner of *Time* magazine's Quality Dealer Award for Toronto.

Recognition is often a direct result of involvement. In the case of John Logan, he is unquestionably involved in his industry and his community. Logan is a member of the Federation of Automobile Dealers Associations and has been its seminar moderator. He has also served as director of the Toronto Automobile Dealers Association, director and president of the GM Dealers Association, and president of the Brampton Automobile Dealers Association. Logan is a Mason and a member of the Scottish Rite, the Brampton Golf Club, and the Brampton Curling Club. He has served as president and director of the Brampton Board of Trade, president and director of the Rotary Club of Brampton, and director of the Peel Memorial Hospital Foundation and chairman of the industrial division of its expansion fund drive. In 1983 Logan was deeply involved in the Ontario Games for the Physically Disabled, and for three years served as chairman for the Arthritis Society's annual fund drive.

John Logan Chevrolet Oldsmobile at one time sponsored the Brampton Junior B "Logan's Chevy's" hockey club, and still sponsors 18 other varied sports teams in Brampton. The company has sponsored two students to Northwood Institute. Logan participates in career day at J.A. Turner High School, and donated a 1983 Chevette to W.J. Fenton High School for shop training. The company believes in returning more than just wages to its supporting community.

Competition has forced some major changes in the industry. Problems are reduced for customers by dramatically upgrading quality control before receiving his or her vehicle, and, when service is needed, by making it pleasant. A customer needing service makes an appointment for diagnosis, drives into the temperature-controlled service reception area, and gets a diagnosis and a computer estimate of time for repair. If the customer waits, Logan provides a cafeteria and full-course meals from breakfast on. At Logan's used car buyers are no longer second-class citizens. They discuss their deals in the relaxed new offices just like new car buyers.

Leasing has made dealers more involved in long-term credit, customer monthly billing, and after-sale maintenance. Being in business for the long term is not a simple catch phrase, it is a necessity brought on by change. The management and staff at John Logan Chevrolet Oldsmobile Inc. are well trained and aware of what they have to do to survive.

PARTNERS IN PROGRESS

The expanded facilities of John Logan Chevrolet Oldsmobile Inc., Brampton's premier General Motors dealership.

DEL/CHARTERS LITHO INC.

Del/Charters Litho Inc. is in the printing business, offering a complete range of printing services, including brochures, P.O.P. posters and banners, annual reports, kit folders, labels, pocket books, newsletters, and news bulletins, to a multinational client list. Crown Life is probably the firm's oldest client, dating back to 1930. Others include Xerox Corporation of Canada, Warner Lambert Pharmaceuticals, the Canadian Auto Workers, and the Federal and Provincial governments.

The town of Brampton was incorporated in 1873. The following year Allen Franklin Campbell began publishing the *Brampton Conservator* newspaper, using a Washington hand press. Located near the four corners, it was a one-man operation.

In August 1890 Samuel Charters took over. The firm prospered under his leadership and continued to grow until the time of his death in 1943. In addition to his business interests, Charters became involved in politics, first running for provincial office in 1903, initially losing to the Liberal incumbent. However, he later served two terms until illness cut his second term short. He also served as mayor of Brampton, and in 1917 won the first of five federal terms that would eventually see him become Conservative Party Whip. Charters retired from politics in 1935, just about the time another young man of 22 was starting his own printing business in downtown Toronto.

Calvin "Kelly" DeLuca, honorary chairman of Del/Charters Litho Inc., started Delgraphics on St. Patrick's Street. Originally a partnership, DeLuca bought out his partner in the late 1960s. The firm's current president, Ronald DeLuca, purchased the enterprise in the mid-1970s after learning the business from his father. The company is now a partnership with vice-president and general manager Jerry Tancredi. Del/Charters Litho Inc. is the amalgamation of these two businesses, owing so much to the individual efforts of Kelly DeLuca and Samuel Charters.

Charters Publishing Company was incorporated in 1919. Ron DeLuca still has the original shares issued at the time. Number one went to Samuel Charters, number two to his son C.V. Charters, number three to J.E. Charters, and number four to S. Wilson.

By 1906 the one-man company had grown to 10 employees. The one newspaper eventually grew to include the *West Toronto Weekly, The Weston Times and Guide, The New Toronto* (and Lakeshore) *Advertiser,* the *Port Credit Weekly,* and the *Guelph Review* before being sold to the Thomson newspaper chain in 1953. In 1943 responsibility for the management of the company passed to Clarence "C.V." Charters, who was followed by Reginald Charters and Samuel Charters II. Finally the shares were sold to Alan H. Charters, who served as vice-president and general manager.

Printed material ready to go to the post office in the early 1920s.

On September 2, 1978, Ron DeLuca and Jerry Tancredi, with funds provided from Delgraphics, purchased the company and changed its name to Charters Litho. Sales increased from $800,000 in 1979 to two million dollars in 1983. The following year Delgraphics and Charters Litho were merged to form Del/Charters Litho Inc., and combined sales exceeded six million dollars.

The firm employed 22 people in 1979 and now employs 50, some of whom have never worked anywhere else. One recent retiree was with the company for more than 50 years, and another was in his 44th year of employment in 1986. The great-grandfather of longtime *Conservator* compositor Harry "Finn" Elliott is reported to have named Brampton after his home of Brampton, Cumberland, England. That was in 1852, when Brampton consisted of 78 people living on two streets, Hurontario and the side road between lots five and six (now Queen Street), and was called Buffy's Corners.

Today the employees of Del/Charters Litho Inc. are all members of one of three unions, but still retain the small-town

The first share issued to Samuel Charters on the incorporation of Charters' publishing company, April 1, 1919.

A new state-of-the-art web press, a Solna Distributor 25, was installed in November 1986.

loyalty that has seen them help the company through tough times. The hands-on management of DeLuca and Tancredi is calculated to bring an aggressive new attitude to an old established business. Service has been accelerated. Extensive new equipment was purchased in 1982, two full-time estimators are on the job, and the company is moving quickly into the computer age.

In late 1986 Del/Charters Litho Inc. installed a modern new web-offset Solna Distributor, a Swedish press—the first of its kind in Canada. The press is expected to offer short make-ready times, virtually maintenance-free operation, and is capable of printing sophisticated color throughout a newspaper. It will be a replacement for a 30-year-old press. The comparison is simple; where the old press could print 14,000 newspapers per hour, the new can print 30,000. With the press installation will come six skilled full-time jobs.

A computer-assisted Heidelberg Speedmaster four-color, sheet-fed press—it runs 12 hours a day, six days a week.

ANTHES UNIVERSAL LIMITED

Anthes Universal Limited markets its office products and art supplies from its corporate headquarters at 341 Heart Lake Road South in Brampton.

Anthes Universal Limited's products are devoted to organizing people, from single individuals with home filing needs, to major corporations needing anything from file folders to antistatic mats for their computer terminals. Anthes Universal is also in the business of providing art supplies, again spanning the complete range of customers from beginning students to professional artists.

Anthes Universal traces its history back to two well-known corporations with long and respected histories in the stationery trade. Office Specialty Manufacturing Co. Ltd. is thought to have started operations in Rochester, New York, about 1888. The firm's first major product was the Shannon Arch File. Its first office was located in Toronto, but by 1898 a vacant factory was purchased in Newmarket, Ontario, producing a line that grew to include filing supplies and wooden and metal office furniture. Office Specialty gained such acclaim as a filing specialist that courses were run in commercial schools granting Office Specialty filing graduation certificates. The firm was acquired in 1961 by Anthes Imperial Limited of St. Catharines, diversifying out of the home heating oil business.

R.J. Copeland of Copeland-Chatterson Limited designed and patented a new security binder in 1891. He established his first plant on Adelaide Street in Toronto in 1896. Copeland-Chatterson became known as the loose-leaf systems specialist of Canada. The firm's growth required an expanded plant, and in 1905 a new factory was built on Railroad Street in Brampton. By the following year the firm had sales offices across Canada. It is reported that at one time the sign welcoming people to Brampton advised, "You are now entering Cope-Chat Territory."

The year 1907 was a year of expansion for the Copeland-Chatterson company. A new manufacturing plant was opened at Dudbridge, Stroud, Gloucestershire, England, which produced a time-saving system called the Paramount Sorting System. The program was effective, selling into Holland, Denmark, Poland, India, and other markets.

During this same year the Rolla L. Crain Co. of Ottawa was acquired, the Crain organization having a considerable amount of printing business with the Federal Government.

A new charter was taken out changing the company name to Copeland-Chatterson-Crain Limited, and this arrangement was in effect for several years. On Christmas Eve, 1915, fire destroyed the new building and equipment housing the Ottawa business. With a war on and the shortage of building materials, the Ottawa business was abandoned. The Crain sales staff merged with the ongoing company and sometime after 1915 a new company charter was taken out in the name of Copeland-Chatterson Limited.

Copeland-Chatterson was noted for innovation and held patents on approximately 170 items, 90 of which were of Canadian origin. Among those were the Kwikway continuous billing system, the Perpetual Purpose invoice system, the Visform visible record system, Style D ledgers, Dual Spirit Duplicators and Appliances, and the Multi-Prong price book system.

In 1961 Anthes Imperial acquired Copeland-Chatterson and moved the Office Specialty filing supplies division to Brampton. That same year Anthes purchased Egry Business Forms Ltd., makers of snap-set and other business forms. The forms portion of the business was located on Judson Street in Mimico. As a result of the various acquisitions, Anthes Business Forms Limited was organized. In 1968 Anthes Imperial Limited was merged with Molson Breweries Limited to become Molson Industries Limited. By 1972 Brampton had become headquarters for Anthes Business Forms Limited, housed in its current plant at 341 Heart Lake Road South.

Molson acquired Haig Printing in 1974 and Willson Office Specialty three years later. In 1979 it sold the Mimico plant to Maclean Hunter and the stationery and office supplies portion of the business to Corvent Developments Limited. When Corvent's owner and president, W.R. Snowden, took over, the firm was known as Anthes Brampton Office Products, noted as Canada's largest manufacturer of file folders.

The innovative Chatterson has been replaced by the equally innovative Wes Snowden. The introduction of the Portafile™ has taken the firm into the home office market, becoming its number one product line. Anthes' "My File" line offers an attractive, colorful line of expanding files. Other popular home filing products include Taxpack, Family File, and Study Mate. A Job Finder kit helps one prepare, present, and mail an impressive resumé.

Among the innovative new directions Anthes has taken since the acquisition is the addition of art supplies. Anthes became the exclusive distributor for the well-known Reeves brand of artists' supplies, as well as the prestigious line of Winsor & Newton favored by top Canadian artists.

In 1981 Anthes broadened its horizons with the acquisition of a U.S. company, Attitype Inc., manufacturer of office machine cushions and dust covers. Under the Attitype brand, new products such as computer dust covers and Static Mats have been introduced.

In 1984 the corporate name was changed to Anthes Universal Limited to reflect its expansion into the U.S. and European markets. The U.S. company, Attitype Inc., became Anthes Universal Inc. In 1985 a plant was opened in Peterborough, England. Anthes U.K. acquired the assets of Page of London, manufacturer of traditional children's metal paint boxes, while Anthes U.S.A. acquired Art Masters Inc., makers of top-quality double-primed stretched and rolled canvas as well as canvas panels and stretcher bars. These products are marketed under the brand name Artists' Choice.

Anthes Universal Limited is a 100-percent Canadian-owned company actively marketing worldwide its office products and art supplies from its corporate headquarters in a modern 100,000-square-foot plant in Brampton. The flair Mr. Copeland demonstrated for designing new and timely products carries forward into the future as the international Anthes team continues to produce new product, anticipate market changes, and offer new leadership today to the office products/stationery/art material world.

The Portafile™ is Anthes Universal's number one product.

A page from a 1953 Copeland-Chatterson Limited catalogue.

BRAMPTON

BRAMALEA LIMITED

One of today's largest real estate companies in North America, Bramalea's grass roots lie in the raw farmland of Peel County. Bramalea Limited retains a lasting identity in the public mind and a close association in ongoing projects with its birthplace, but the company bears little resemblance, except in spirit, to the framework of forerunner Bramalea Consolidated Developments Limited.

While the original alliance of Canadian and British interests was concerned

The management of Bramalea Limited is vested in two chief executive officers, vice-chairman Benjamin Swirsky (left) and president Kenneth E. Field.

solely with the creation of the country's first satellite city—a formidable challenge in itself—the contemporary Bramalea reaches to the skylines of Texas and the coastlines of British Columbia and California. Today Bramalea Limited, although remaining a major home builder and community developer, manages a $4-billion portfolio of income-producing properties comprised of shopping centres, office buildings, business parks, hotels, and residential apartment buildings.

The New York Life office building is one of 13, totalling 3.5 million square feet in all, owned and managed by Bramalea Limited in the Metropolitan Toronto region.

Records dating to the incorporation of Bramalea Consolidated Developments Limited in Ontario on December 11, 1957, reveal that an abundance of vision and perception was shared by the founding members. The bold concept of a "new town" between the established community of Brampton and northwest Metropolitan Toronto was first pursued by a Canadian company, Bayton Holdings Limited of Toronto. In a 2 1/2-year period prior to March 1958, at the direction of Bayton principals Dr. J.C.T. Sihler, G.D. Clarke, and R.C.C. Henson, an initial area of 5,085 acres was assembled under agreement with 44 farmers at a total cost of approximately $5.5 million.

This tract, then a part of the former Township of Chinguacousy, was the largest single block of land (7.94 square miles) ever assembled for the purpose of erecting and adjoining a satellite city to a Canadian metropolis. Subsequently, as the venture progressed steadily, the Bramalea area would be increased to 6,365 acres in the 1960s and to more than 8,000 acres in the 1970s. Throughout the rise of Bramalea, from early relationships with the township and after 1974 as part of the City of Brampton in Peel Region, mutual co-operation would prevail between the company and municipal governments.

In 1958 a syndicate of three British firms (Eagle Star Insurance Co., Cayzer Trust, and British Electric Traction), under the direction of Close Brothers Limited of London, England, a merchant banker, was introduced to Bramalea. Additional capital was contributed to Bramalea by these firms, which then purchased the assets of Bayton. An eventual expenditure of $50 million by Bramalea Consolidated Developments and "hundreds of millions more" by builders, industries, and other interests was forecast over the ensuing decade.

Land preparation, installation of

services and roads, and construction led to the completion, in 1959, of the first 300 houses in Phase One of Bramalea between concessions 3 and 4 south of Highway 7. The first family moved into one of the residences, which were priced from $12,000 to $16,900, in the spring of 1960. The first industry, Northern Electric Co. Ltd., arrived in 1961 to occupy a 735,000-square-foot plant on a 100-acre site, built at a cost of $16 million. It is important to note that Bramalea generally averted the plight of other municipalities, which suffered financially from a heavy load of residential assessments, by attracting industry on a balancing scale in the formative years. By 1972 the ratio was a healthy 47-percent industrial to 53-percent residential.

The launch of Bramalea, and its emergence as a viable and profitable urban entity, was not without serious problems arising from external economic and political events. The firm sustained a body blow of near-fatal consequences from the February 1959 decision by Prime Minister John G. Diefenbaker to scrap a multimillion-dollar national defence commitment to the CF-105 (Avro) Arrow. Even though Bramalea was founded on the basis of consultation with leading realtors, architects, and developers, as well as official planning projections of long-term need for new homes and industries on the Toronto perimeter, the proximity of the giant A.V. Roe and Orenda Engines aircraft assemblies and their 15,000 employees at Malton was critical to the inception of the new town.

However, Bramalea survived that immediate crisis, just as it later rode out periods of high interest rates and economic decline. If there exists an operational armature responsible for the ultimate success of the corporation, it reposes in the resiliency, resourcefulness, and unwavering determination brought to bear in times of adversity.

The Bramalea City Centre, constructed in 1973, is one of Bramalea's flagship centres and one of the largest in North America.

More than 350 companies have located in the Bramalea Business Park since development began with the creation of the Bramalea community.

An integral business and social component of the City Centre, Bramalea's Holiday Inn Holidome hosts both business travellers and local visitors.

Since constructing the first homes in the Bramalea community, Bramalea Limited has earned a reputation for quality design and construction of houses and condominiums in southern Ontario and California.

From the outset Bramalea Consolidated Developments held fast to a firm commitment to participate in the transition of up to one million people to the outer ring of Toronto suburbs and the eventual requirement of some 32 square miles of industrial sites by 1980, according to the 1957 Royal Commission on Canada's Economic Prospects. Within the satellite city, the company proceeded vigorously on all fronts—sales and administration, public utilities and services, schools, churches, housing, roads, building supplies, and even landscaping centres. From gas stations to the Civic Centre, the undertaking was totally planned under strict controls in a mode of master design drawn from the postwar British experience.

The early 1960s saw Bramalea turn the corner, commencing in 1962 with a base population of 4,000 in 1,000 residences, an upswing in industrial site sales, and completion of the first composite secondary school. At the same time plans were unveiled to introduce town houses (at $13,000 per unit) and apartment buildings, in effect bringing about a much broader range of accommodation. Ground was also broken for the first small shopping centre, comprising nine stores, while athletic fields, parks, and community services such as a medical-dental centre were provided.

The progression of low-rise and high-rise residential buildings proceeds apace today, offering housing to every income group in single and multiple-type structures, including outright ownership, condominiums, and rental arrangements. Today the Bramalea sector of the City of Brampton has a population approaching 90,000, and the Bramalea Business Park is home to more than 300 manufacturing, distribution, and technical firms, and still counting.

Bramalea Limited traces its substantial buildup of holdings in the continental property market to an early propensity for diversification. The acquisition in 1963 of the Niagara Peninsula Shopping Centre, a 52-store assembly on 60 acres in south St. Catharines, and the smaller Applewood Village Shopping Centre (sold in the late 1960s) on the Queen Elizabeth Highway in Toronto Township formed the beginning of an ambitious program of widespread investment across Canada and the United States. Experience in the retail-commercial field was also applied to the development of Bramalea City Centre as a focal point for peripheral hotel, apartment, and service amenities in that locale.

By 1983 the Bramalea City and Pen centres had grown to 1.2 and 1.0 million square feet, respectively, to command recognition as flagships of the company's vast retail interests. Today plans are being finalized to expand, renovate, and re-merchandise both centres to ensure their supremacy within the shopping centre sector.

The magnitude of the shopping centre holding is illustrated by the creation in

1986 of Trilea Centres Inc., a new Bramalea subsidiary, to own and manage the 30 Canadian shopping centres formerly owned by Bramalea and Trizec Corporation Ltd., Bramalea's principal shareholder. With a book value of $1.4 billion and leasable space exceeding 13 million square feet, Trilea ranks as one of the largest consumer sales groups in North America. The numbers are even more impressive when Bramalea's U.S. division of 20 shopping centres and more than five million square feet is added to the roster.

While the firm widened horizons and investments throughout the 1960s and early 1970s, the Bramalea community continued to fulfill the promise of the founders. Concurrently, the flourishing proprietor pursued new projects in Western Canada—at Vancouver, Calgary, and Edmonton—and in a number of southern Ontario cities.

Those were heady days for the corporation—a metropolitan base, assets exceeding $150 million, the Bramalea City Centre a fait accompli, and land approvals flowing in from Chinguacousy and other Canadian centres. The higher profile surfaced in the 1973 annual report, and within two years the corporation would again shift gears with a reorganization of top management and a repatriation of the former English ownership to all-Canadian control.

Today the company operates under

Jaguar Canada Inc. is one of many major companies that have established their corporate offices in the Bramalea Business Park.

the entrepreneurial direction of Kenneth E. Field and Benjamin Swirsky, the joint chief executive officers. The whole exercise was crowned in 1976 by a change in name to Bramalea Limited, recognizing maturity as a comprehensive real estate organization transcending national boundaries.

Buoyed by the most successful year in the firm's history in 1978, with profits topping $9 million and assets of $400 million, Bramalea Limited embarked on an aggressive residential and industrial quest in the southern United States. Initial realty investments at Boca Raton, Florida, and Los Angeles, California, set the stage for further involvement in key growth cities such as Dallas, Texas, and Denver, Colorado, all of which contributed substantially in 1980 to an overall increase of $250 million in new assets.

The effects of restructuring, the spread of investment, and renewed emphasis on prudent, judicious management were most apparent during the calamitous economic recession of 1981-1982. Bramalea was positioned by that time to not only survive the worse slump since the Depression, but also to successfully launch the billion-dollar Dallas Main Center in the heart of that vibrant city. The first phase of the project, the magnificent First Republic-

Bank Plaza, a 72-storey building with 1.8 million square feet of leaseable space, has now been completed.

In the wake of the recovery in 1984 the assets of Bramalea Limited soared beyond the $2-billion plateau. Almost simultaneously, American Motors/Jeep/Renault announced a $764-million assembly plant on 167 acres purchased from Bramalea in the Bramalea Industrial Park, thereby adding another dimension of affirmation to the decision taken 30 years ago in rural Chinguacousy Township. And just last year the firm completed the acquisition of Fidinam Properties to fold in an additional $450 million worth of real estate.

By every measure of attainment Bramalea Limited qualifies as a signal chapter in the annals of Canadian business history, as well as an inseparable and motivating force in the rising fortunes of the City of Brampton.

Bramalea Limited initiated development of apartment projects in the early 1960s and today owns and manages 15 high-rise rental apartment buildings in the Bramalea community.

THE RICE GROUP

LEFT: Louis A. Rice, president.

RIGHT: Maxwell C. Rice, vice-president/secretary.

The Rice Group represents the current phase in the evolution of Rice Construction, builders of family subdivisions as well as industrial and commercial properties. Rice is a leading developer of adult life-style communities as alternatives to simple tract housing. Typical examples are preretirement homes for still-active empty-nesters, and first-home communities with smaller but expandable houses built on lots large enough to allow for future additions.

Lou's daughter Dyanne and Max's sons, Rod, Jeff, and Dave, all active members of The Rice Group today, represent the third generation of the Rice family to be involved in the business. Their grandfather and great-uncle emigrated to Canada from Newfoundland before Confederation and between the first and second world wars, working as carpenters in the Toronto area. Rod's father, Max, and two uncles, Tom and Lou, became carpenters also; when they returned from wartime duties, they founded Rice Construction in the late 1940s. The first offices were at Jane and Wilson in Downsview above the hardware store, which was owned by another uncle, Gord. At that time most of the homes built by the firm were in the Downsview and North Toronto areas.

In 1954 the Toronto market slowed, and, in what was originally intended as a temporary move, Rice undertook its first combined residential and industrial project in Brampton. Liking the outcome, the company remains in Brampton and is positive about the future of the city. That first local project consisted of 700 homes in the Eldamar subdivision near Highway 10 south of Clarence, as well as industrial and commercial property along Kennedy Road.

In the 1970s Rice established three empty-nester communities. Sandycove Acres south of Barrie, Grand Cove Estates near London, and Wilmot Creek east of Oshawa provide 3,000 single-family houses, privately owned but built on land leased from Rice, which provides a wide range of social and recreational amenities in these communities.

Current Brampton projects include Rice Business Centre, with 23 acres and 300,000 square feet of available property in the northwest corner of town; the 140-acre Rice Industrial Park, located east of Airport Road and north of Highway 7; and a joint commercial and residential development with Kerbel Developments in the corridor south of Steeles Avenue and Highway 10.

BERGMAN GRAPHICS LIMITED

The array of electronic systems—consoles tapping instructions for high-resolution color, copy alignment and alteration, and ultimately page assembly for the printing industry client—is light years advanced from the initiation of Gerald Bergman into the fascinating world of lithography. Yet Bergman's apprenticeship as a camera operator and film stripper in Toronto area shops during the early 1960s formed the basis of a career decision and a solid foundation for his success in developing one of the leading graphic arts companies in Ontario.

Bergman Graphics Limited, a modern plant built on one acre of land at 26 Bramsteele Road in 1973, houses high-tech equipment capable of scanning and composition functions for the pre-press market unheard of when Bergman entered the trade in 1961. Laborious, time-consuming procedures have been replaced by sophisticated film separation and reproduction by fingertip command. During a recent five-year upgrading program, the firm became the first in Canada to install a laser color scanner and a fully electronic stripping system.

Bergman's marriage to Marlene Huggins of Brampton in 1964 and a move to the city two years later set the scene for the incorporation of Bergman Graphics in February 1967 to meet a demand for film and type from local printers. Today the couple are co-owners of the company and have two school-age children, Steven and Peggy. The family is active in community and church affairs, with Gerald Bergman serving on the boards of Kennedy Road Tabernacle and Christian School.

Bergman Graphics' first shop was located in the basement of the family home on McMurchy Street, where volume soon grew to require one full-time and two part-time employees. In less than four years a store basement was rented as the staff expanded to include five full-time and two part-time positions. Upon moving into the present building, the Bergmans bought out a small graphic arts firm with a good clientele, which combined with a steady increase in orders to require 17 full-time employees by 1977.

The permanent staff of Bergman Graphics Limited in 1987 totals 50 people, who prepare all of the art for such valued accounts as the Miracle Mart national flyer and Trivial Pursuit board games. In February 1987 the company capped 20 years of successful operation by opening a fully equipped division in Burlington, Vermont.

Gerald Bergman, founder.

HUEGENOT LIMITED

Huegenot Limited operates as a privately owned, full-service wholesale food distributor. Primarily it services the Burger King Canada Inc. national chain of restaurants, but is unique in that it also services assorted federal and provincial hospitals, institutions, and armed forces bases. In addition to supplying clients with frozen, refrigerated, and dry food products, Huegenot provides a line of paper products and cleaning supplies.

Huegenot Limited began operations in March 1976, servicing a number of restaurants and fast-food outlets in the Kingston area of Ontario. The business started with a small warehouse, one truck, and five employees—four of whom are still with the firm. From Kingston the business quickly expanded into Eastern Ontario, and later into the Toronto and Windsor areas.

In December 1977 Burger King recognized the handling and health advantages of changing from a fresh to a frozen meat supply. This switchover allowed Huegenot to become a full-service distributor on a much wider scale, covering southern Ontario and Quebec. With the opening of its first Toronto warehouse in March 1980, distribution was expanded to northern Ontario and Winnipeg.

Two more warehouses were opened in 1982. From Winnipeg, distribution was extended to Calgary, and from Moncton, the eastern provinces were included. At the same time a search was started to find a site close to main transportation routes and suppliers, while avoiding high-density work areas and industries, and, importantly, meeting the community and social expectations of Huegenot employees. The city of Brampton met or exceeded all requirements, and on April 16, 1984, operations began at its present location at 1310 Steeles Avenue East and Dixie Road. Shortly thereafter the decision was made to move the head office to Brampton, relocating most of the senior management people and their families. Today 80 percent of the Brampton staff live in the city.

A new warehouse facility in Calgary was opened in June 1985, and by December of that year Huegenot had become a national distributor by extending service to British Columbia. In January 1986 the Equipment Distribution Services Division was formed to provide a full line of restaurant equipment. Huegenot is already planning future expansion.

Today a staff of 100 operates out of Huegenot Limited's warehouse centres in Calgary, Winnipeg, Toronto, and Moncton, directing a fleet of company-owned multitemperature trailers and servicing customers nationwide on a weekly basis.

Huegenot Limited, a full-service wholesale food distributor, chose Brampton as the ideal location, and the head office and warehouse were moved there in 1984.

Huegenot Limited began operations in this head office and warehouse in Kingston. Photo circa 1981

KORD PRODUCTS LTD.

Kord Products Ltd. is in the business of providing containers and packaging for the horticultural, agricultural, and medical laboratory industries. Just prior to locating a satellite plant in South Carolina, principals of the firm were asked by one of the directors of the local chamber of commerce what interest there would be in a U.S.-based Kord facility. The chamber representative was told to go home and check all the flower pots being used there. Of the 19 plant containers, 17 had been manufactured by Kord.

The company was started in 1968 by its current president, Ralph Welsh, in a small Cooksville plant near Dixie Road and Dundas Street. There the six-member staff produced a line of custom laboratory equipment using moulds provided by customers. Today Kord Products Ltd. still supplies a substantial volume of Petri dishes sold and distributed across North America.

In 1973 the firm moved to Orenda Road in Brampton because the area provided a desirable "quality of life" environment for employees. The company purchased the former J.E. Searle plant, a 17,000-square-foot facility on seven acres of land. The plant was expanded to 37,000 square feet before Kord's occupation, and another 10,000 square feet were added after the first year. The year 1979 brought a 58,000-square-foot addition, and the current expansion was completed with a 9,000-square-foot mezzanine. Future needs will be provided for by the use of property just north of the existing plant.

The Kord product line has broadened over the years through a series of acquisitions. In 1970 a small injection-moulding company was added and moved from Bradford to Cooksville. The firm produced items as diverse as golf bag bottoms and flower pots. When its major distributor of flower pots went bankrupt, Kord Products Ltd. attacked the market directly and became a major player in the flower pot trade. The sizing standard used by Kord is now accepted as the industry standard, and the firm was one of the first to use metric sizing.

In 1973 the company purchased a fibre operation in Burlington that is now called Kord Fibre Products Division. At about the same time the Brampton plant introduced vacuum forming for bedding plant trays and expanded its injection-moulding process. Kord Corp. was established as a sister plant in 1981, operating from Lugoff, South Carolina; Kord Erin Ltd. was opened a year later in Ireland.

By 1986 Kord Products Ltd. employed more than 350 people across North America. The plant operates seven days per week, offering employment to students, often from age 17 through university, some of whom eventually become full-time staff members.

Kord Products Ltd., located on seven acres at 390 Orenda Road, manufactures containers and packaging for the horticultural, agricultural, and medical laboratory industries.

BRAMPTON HYDRO-ELECTRIC COMMISSION

The history of Brampton Hydro-Electric Commission is typical of the history of many municipal utilities in Ontario.

Huttonville, or Huttonsville, was named after James Patterson Hutton, who established a lumber mill in the area about 1850. His son, James Oscar Hutton, built a shoddy mill next to the lumber mill to manufacture reclaimed wool fibres into new material, powering the plant with a Thompson-Houston dynamo generator. Using a 2,200-volt transmission line, he introduced electric light to Brampton with a single carbon-arc lamp in front of the Queen's Hotel on Queen Street in 1886. The light was so brilliant it allowed people as far away as the Grand Trunk Crossing to read the fine print in the newspaper.

The town adopted a plan to install 18 streetlights for street illumination at a cost of $72.22 per lamp per year on the understanding the lamps were to be turned off after 12:30 a.m. nightly, as well as all night on evenings preceding a full moon and for seven nights after, except when clouds hid the moonlight. In 1903 the customer base increased from 43 to 500, and included W.B. McCulloch's planing mill, the first industrial user of power in Brampton.

In 1906 the Hydro-Electric Power Commission of Ontario was formed to harness Niagara; five years later Brampton hooked into the system through an 11.9-mile line of 510 poles running from Port Credit. The system went into continuous service on October 16, handling a peak load of 21 horsepower. By 1961 Brampton Hydro's peak load was 28,500 horsepower, and in 1986 that figure had reached 227,000 horsepower.

In 1912 the Hydro-Electric Power Commission bought out the Hutton source and became the sole supplier of electricity to Brampton. Six years later the city's streets became lit all night, and in 1923 electric service was brought to a 60-square-mile rural area outside of Brampton, fulfilling the *Conservator's* prediction of December 15, 1910, "that . . . hydro electricity will transform work on the farm, most gratefully the work of women, from drudgery to a healthful and pleasureable occupation."

The first Hydro-Electric Commission of Brampton was comprised of J.S. Beck, chairman; B.F. Justin, commissioner; and Mayor T.W. Duggan. The staff consisted of G. Ostrander, superintendent, and a Miss Sheppard as secretary. Brampton Hydro's first home was in the Heggie

Brampton, circa 1911, looking west on Queen Street before the hydro wires were placed underground.

Electricity, generated by a Thompson-Houston dynamo generator housed in this James Oscar Hutton-built shoddy mill, first came to Brampton in 1886.

This October 1949 photo shows Hydro men removing a washing machine from a home to convert it to Ontario Hydro's new 60-cycle frequency.

In the 1980s hydraulic equipment is used extensively in overhead line work. This pole and its associated equipment has been installed with the conductors still energized.

Building on Main Street South. After the severe power shortages of 1948, the commission authorized the conversion of the complete power system from 25 to 60 cycle. Over five years, and at a cost of two million dollars, the electricity was turned off one street at a time, and electricians would go through each individual house converting every electric appliance to the new frequency.

In 1955 the three-member staff moved south one building on Main Street, sharing space with the water commission. At that time consumption of electricity was 30 million kilowatt hours; by 1985 that figure had grown to 1.653 billion kilowatt hours. In 1965 the commission moved into spacious electrically heated and air-conditioned offices at 6 George Street South, which even boasted a soundproof room for the second-hand billing machine purchased in 1953 to eliminate manual processing.

A move was made in 1978 to its current headquarters at 50 Main Street South. On January 1 of that year a new citywide Brampton Hydro-Electric Commission expanded the territory it served from 16 to 96 square miles, increasing its customer base from 14,000 to 31,000. The effect of this rationalization of the power

RIGHT: A road gang of the 1890s on Queen Street East in front of the Queen's Hotel.

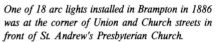

One of 18 arc lights installed in Brampton in 1886 was at the corner of Union and Church streets in front of St. Andrew's Presbyterian Church.

based automatic or semi-automatic control system enables an operator, from a single consul, to pinpoint a problem, isolate the fault, and reroute power around the area with a minimum of power disruption. The SCADA system is also used for monitoring loads, substations, and line equipment. The K.R. Taylor Service Centre on Glidden Road houses both the control room for the SCADA system as well as the planning, design, operations, and construction departments.

By 1986 Brampton Hydro employed 173 people. They run a fleet of 67 trucks, 18 of which are ecologically efficient propane-powered vehicles. There are now 13,112 streetlights and 987 park walkway lights in the City of Brampton.

Brampton Hydro-Electric Commission has grown dramatically with the City of Brampton. By May 1986 the utility was serving more than 50,000 customers, spread over 96 square miles. However, it is important to note that Brampton Hydro has not lost its awareness of aesthetics in the wake of this boom in business. In 1986 the installation of coach-style streetlights was completed downtown. They may not be a true replica of J.O. Hutton's arc lamps, but they are a reminder in spirit of the city's long electrical heritage.

LEFT: It is assumed that this flour bag denoting "electric" was a result of the availability of electricity for grinding.

needs in the Peel Region made Brampton Hydro the 10th-largest electric utility in Ontario by 1983.

The installation of a Supervisory Control and Data Acquisition (SCADA) system began in 1980. This computer-

BELOW: Main and Nelson in the 1890s, with an arc light just visible in the upper left hand corner.

LAWRENCE LAWRENCE STEVENSON
Barristers & Solicitors

Sixty-three years ago classmates and close friends Harold R. Lawrence and Gordon Graydon graduated from Osgoode Hall and, with youthful ambition, opened their own law office in Brampton.

The year was 1924 and Brampton, with a population of 4,551, was the centre of a prosperous agricultural county. Local dairy herds were renowned and, with Dale's roses being sold throughout North America, Brampton was truly the Flowertown of Canada. And in the thriving two-man law office of Graydon and Lawrence, the secretary dutifully pounded out the wills and real estate transactions on the ponderous black Smith Corona.

Once a week one of the partners would set out on the dusty rural roads and journey to their branch offices in Bolton, Palgrave, Malton, and Streetsville—often providing the only legal service in the community.

In the 1930s Gordon Graydon heeded the call to politics and was elected as the federal Conservative member for Peel, becoming acting House Leader through much of World War II. At war's end he made national history as an original signatory to the United Nations Charter on behalf of Canada. Closer to home, partner Harold Lawrence continued the tradition of public service and was elected mayor of Brampton from 1949 to 1951.

In the mid-1950s the law firm saw history repeat itself when Harold Lawrence's son, William, joined the practice. In turn he was later joined by his law school classmate and friend, Basil Stevenson, and the firm's name was changed to that by which it is known to the present day—Lawrence Lawrence Stevenson.

Today this long-established law firm has become the largest in Brampton, with 17 lawyers who provide seasoned expertise in corporate and commercial law, land development, civil litigation, family law, and, of course, as tradition would have it, wills and real estate transactions. It's a law firm that is as contemporary as its state-of-the-art computers and legal practice systems, which allow it to stay on the leading edge of legal practice in a rapidly changing legal world.

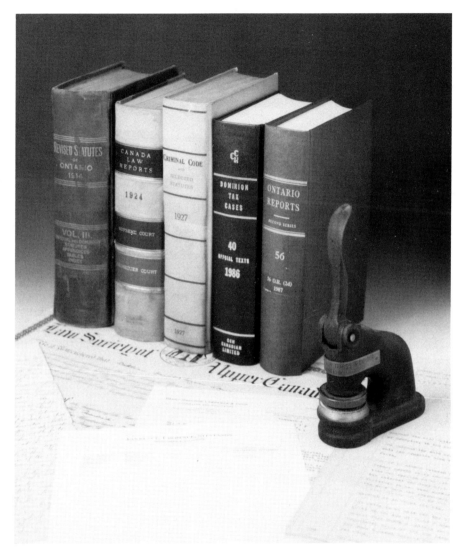

And it is a unique law firm with a deep loyalty and civic commitment to the City of Brampton—a loyalty that stems from 63 years of full involvement and active participation in community affairs. Partners of the firm have served as chairman of the Peel Memorial Hospital Board, city mayors, as well as long-term members of service clubs, sports organizations, and church groups. The current president of the Brampton Board of Trade is Dennis Cole, partner in Lawrence Lawrence Stevenson.

The history and growth of Lawrence Lawrence Stevenson has paralleled the city's own prosperity. Both may have started out on a small scale, but both have shown tenacity and spirit, and have flourished far beyond expectations of the founders. Together we all look forward to a future of even greater success and prosperity.

EXPORT PACKERS COMPANY LIMITED

Export Packers Company Limited has evolved over 50 years from a modest, small food retailer into a significant factor in the Canadian food-processing, wholesaling, and biotechnology industries, and has become a major exporter of food products and pharmaceuticals to four continents. The firm supplies three distinct product groups: frozen meats, poultry, and seafoods; egg products; and biotechnical products.

In frozen foods, Export Packers distributes to prominent Canadian and U.S. wholesalers bulk frozen meats, poultry,

The Rubenstein family in 1926. Father Herry is second from right and mother Edith second from left, with children (from left) Fay, Toby, Max, and Rosie.

Max Rubenstein in 1940, three years after the founding of Export Packers.

and seafoods under private label or under the Family Delight brand, a trademark of Export Packers. The Winnipeg plant breaks and processes 1.2 million eggs per day on a single shift. The eggs are frozen, dried, or liquified as either egg whites, yolks, or whole eggs. They are then sold to the baking, dairy, and food-processing industries; specialty food processors; and the hospitality trade, which turn them into pies, pastries, pasta, breads, soups, ice cream, cookies, cakes, and yogurt.

Export Packers is the largest processed egg supplier to the Canadian market, and is increasing its market share in Europe, the United States, Central America, and South America. The biotechnical products are developed from the 17 proteins found in the egg. The company is now the world's largest producer of lysozyme, effective as both an antiviral agent in the treatment of hepatitis, shingles, and herpes, and as an alternative to nitrates in the processing of cheese and the preservation of meat.

The history of Export Packers is really the story of Max Rubenstein, chairman of the board. In 1926 his father, Herry, borrowed $50 from Mr. Dunkelman of Tip Top Tailors in Toronto, with which he paid the first month's rent on a kosher retail poultry store with living quarters above at Dundas and Euclid streets in Toronto. Herry's father owed $3,000 to the United Farmers Co-operative at Adelaide and George streets. Herry guaranteed the loan in exchange for three crates of chickens, which became his first inventory. He vowed that one day he would own the United Farmers plant. Within 10 months the loans had been repaid, and the retail store eventually was selling chickens to 2,000 customers. By 1932 the retail poultry store expanded and moved to College and Clinton streets.

When Max Rubenstein was 15, Herry asked him to take a look at the bookkeeper's accounts. They were in disarray, but Max corrected them in a single night. The next day Herry took him out of high school and put him to work full time. Export Packers was founded in 1937, when Max Rubenstein, by then 16 years of age, hired three salesmen to solicit the hotel, restaurant, and retail butcher shop business. New refrigeration systems were installed enabling them to compete in the wholesale distribution trade.

The firm's first poultry-killing plant, in Brussels, Ontario, expanded the supply sufficiently to take on chain stores while still holding the share of business from restaurants and butchers. As poultry supply became a problem for the entire industry, Canada Packers and Export Packers became the first companies to import poultry from the United States. Export Packers became and still is one of the largest poultry importers in Canada.

In the mid-1940s the Manning Avenue poultry plant was acquired and a section of it was converted to the production of liquid eggs. In 1955 another goal was realized—the United Farmers Cooperative plant was purchased and

PARTNERS IN PROGRESS

converted to an egg-breaking operation. Export also acquired a plant in Dublin, Ontario, which was a creamery and egg-grading operation.

What is now the Winnipeg egg-processing facility was founded by another father-and-son partnership that sold out to the Ogilvie Mills people. They in turn sold to Labatts, who later sold to Export Packers. The operation was located in leased quarters in a storage plant until, with the help of the Federal Department of Regional Industrial Expansion, a modern, 65,000-square-foot facility was built in Winnipeg.

Fifty years after its inception, the small business Herry and Max Rubenstein started in Toronto employs approximately 140 people in Brampton and Winnipeg. The staff includes Rubenstein's two sons, Jeff, who is chief executive officer, and David, director of the seafood division, and John H. Lee, president and chief operating officer, all of whom have hands-on experience in the business. Export Packers also owns Family Delight Foods Incorporated in the United States. In addition, the firm has a 50-percent interest in Kwinter Packers Limited in Toronto and SMC Meat Services Limited in Vancouver. Consolidated sales exceeded $130 million in fiscal 1987.

While the food portion of the business is expected to continue to expand, Export Packers is making a definite move into the high-tech area to complement its food commodities. More research is being done to utilize the enzymes available from eggs. The company believes that eventually more eggs will be broken for science than for food.

Each year Gold Medallions are awarded to Canadian firms for exceptional accomplishments. On October 9, 1986, Export Packers Company Limited of Brampton was awarded a Gold Medallion as the first-place winner of the Canada Awards for Excellence in the Technology Transfer category. The award recognized the firm's work in successfully adapting a European process for enzyme extraction to the Canadian egg-processing industry.

Max Rubenstein in 1986.

Export Packers Company Limited, Processed Egg Division, at 70 Irene Street in Winnipeg.

GRAFF DIAMOND PRODUCTS LIMITED

Cutting and drilling concrete with diamond-tipped tools is now a common-day event within the construction industry. However, over 20 years ago this method was still in its infancy, and that is when Donald Edward Graff had a dream to create a company that not only provided the contracting services of concrete drilling and sawing, but also manufactured the tools needed—the diamond core bits and saw blades. Since that time the industry has grown rapidly, and Graff Diamond Products Limited has become one of Canada's largest manufacturers of construction diamond tools, with sales reaching across North America. Its contracting division, working mostly within Metropolitan Toronto, provides all of the services and expertise needed for concrete sawing and drilling on a contract basis.

Graff Diamond Products Limited opened its first facility in Weston, Ontario, but has been headquartered in Brampton since 1972. Located at 35 Hale Road, in a 10,000-square-foot plant, the company has grown steadily each year and presently employs more than 60 people.

Since Mr. Graff's death in 1985, the firm has been managed equally by his three children. Richard Graff is president, while Don Graff, Jr., oversees the construction manufacturing division. Daughter Marianne Vardon manages the industrial tool division, which manufactures diamond wheels for industrial grinding. The contracting division, which provides the drilling and sawing services, is managed by Steven Ewles, who started with the firm over 18 years ago.

The company still enjoys the benefits of a North American patent obtained by Donald Edward Graff, Sr., for the Graff Curb Cutter, a machine that revolutionized the method by which curbs are removed. If a driveway opening is needed where a curb already exists, the old procedure would be to jackhammer and replace the concrete, a time-consuming and expensive method. With the Graff Curb Cutter, the curb is literally sliced away with a diamond blade, quickly and efficiently, and left for immediate use.

Graff saw his dream become a reality, and as Graff Diamond Products Limited continues to grow there is no doubt that he would be pleased and proud of the company that bears his name.

Donald Edward Graff, Sr., founder.

One of the many Graff vehicles servicing the Toronto area.

The Brampton head office and manufacturing plant of Graff Diamond Products Limited.

KNECHT AND BERCHTOLD LTD.

Knecht and Berchtold Ltd. is a leader in the landscape industry. Its Alpha Precast subsidiary manufactures a line of concrete site furnishings that includes park benches, picnic tables, concrete planters, trash receptacles, bicycle racks, games tables, drinking fountains for the handicapped, and telephone booths. The firm uses a variety of aggregates to match existing buildings, while providing a product capable of surviving extremes in temperature as well as being relatively impervious to vandalism, damage, or theft.

Alfred Knecht and John Berchtold studied landscape design and construction in their native Switzerland. They later served lengthy landscaping apprenticeships that led to master's degrees. The pair emigrated to Canada intending to head for British Columbia's mountains, but a stop in Brampton convinced them to settle there instead. A partnership was registered in 1953, and the business began with the aid of Tindale Rutherford, who owned a farm on the Second Line East (now Heart Lake Road).

In 1954 Berchtold returned to Switzerland to marry Ruth Baumberger, currently secretary/treasurer of the firm. The business operated out of the Berchtold home in Brampton throughout the 1950s until its present facility at 32 Rutherford Road South was purchased in 1964. Alpha Precast actually got its start in the basement

This facility, at 32 Rutherford Road South, has been the home of Alpha Precast since 1964.

of the Berchtold home in 1958, but did not become a separate entity until six years later.

Knecht and Berchtold was incorporated in 1962, and the partnership continued until 1983, when Knecht retired and Berchtold purchased his partner's share of the company. What began with two ambitious young men in 1953 has grown to a core of 40 employees, with up to 55 people employed during the spring, summer, and fall. In 1986 the firm recorded sales approaching three million dollars.

The second generation of both families is well represented in the landscape industry today. Peter Berchtold manages production of Alpha Precast, and Ron Berchtold is involved in landscape design and construction. Thomas Knecht works for the company on a part-time basis, Steven Knecht has worked there during

The landscape firm's first job was for the Biggs family on Etobicoke Drive in 1953. At that time the company specialized in natural stonework, which required considerable skill. Thirty-two years later the Biggs family still enjoys the stonework and their garden.

summers, and Marcus Knecht joined a Toronto firm after studying landscape architecture at the University of Toronto.

The popularity of Alpha Precast products takes the finished product to many locations. In Brampton, the company's work can be seen in front of the Consumer's Gas Building on Queen Street, the new Bartley's Square Mall, the Four Corners, and in the parks. The City of Toronto uses Alpha site furniture and custom precasts in streetscapes, public squares, and parks. Other examples can be found in Las Vegas, Hawaii, Saudi Arabia, and even in Russia. Knecht and Berchtold Ltd. is a definite leader in its field.

DIECO MANUFACTURING LTD.

Dieco Manufacturing Ltd. specializes in service to the steel tubing industry and is probably the only company in Canada that limits itself to this industry. The firm builds custom specialty equipment. Dieco engineers will visit a customer's plant, assess the problem, then, taking into account the customer's needs, take a machine from concept to engineering, through manufacturing to installation and supervision of early use.

Dieco Sales Ltd. is strictly a sales subsidiary, providing standard machinery from manufacturers in Italy, Germany, and England. These machines are designed to cut, form, and bend steel tubing. Dieco Sales has exclusive sales rights across Canada for three major European manufacturers. Surprisingly, neither firm has ever had a salesman on the road; all work and leads are from referrals or old customers.

The company's president, Dieter Schmitt, emigrated to Canada from Germany in 1960 at the age of 18. A trained tool and die maker, he intended originally to visit for only one year. In 1965 Schmitt founded Dieco Manufacturing. The firm's first plant was in a 400-square-foot former World War I munitions factory in Georgetown. By 1966 the firm had moved to a rented, 2,000-square-foot facility on Queen Street where Queen Street Electric is today. In 1971 expansion and added equipment required a move to a 3,000-square-foot plant on Orenda Road. In 1976 Schmitt purchased another company in Mississauga and opened an engineering office.

The big breakthrough came in 1978, with two large contracts to produce materials-handling equipment for a U.S. firm. That same year Dieco Manufacturing Ltd. moved into a 10,000-square-foot facility at its present location at 67 Selby Road. Within eight months the company added another 8,000 square feet and consolidated the fabricating, manufacturing, and engineering divisions. Today Dieco can handle all aspects of construction in-house, and is anticipating further expansion.

Dieco currently employs 32 people, all of whom Schmitt believes contribute directly to the success of the company. "Growth and success would not be possible without loyal and effective employees," he says.

Today Dieco Manufacturing Ltd., which currently designs and builds automated work cells and computerized machines, is moving swiftly into more sophisticated equipment. The auto industry, for example, is demanding just-in-time deliveries. Dieco is meeting that need with a system it is going to patent that allows a manufacturer of auto seat frames to change over from one maker's frame to another in 30 seconds to 15 minutes, instead of the normal six hours.

PARTNERS IN PROGRESS

JOHN THURSTON MACHINE LIMITED

John Thurston Machine Limited is a highly diversified organization, offering a complete range of metalworking services to industrial customers. Approximately 50 percent of the business is service work: repairing equipment, installing new parts either from inventory or made in-house to suit the need, moving, and installing complex machinery and machinery lines. Another part of the business consists of manufacturing special machinery to suit a customer's specific need.

The company is also a custom manufacturer supplying fabrication and machined parts such as tanks, conveyors, shafts, bearings, weldments, and machining parts to local industry. It also supplies fabrications and machined parts, including fan casings, vibrator hoppers, hubs, and rollers, to original equipment manufacturers on a continuous production basis.

The business was started in 1962 by Len Thurston, father of current owner and president John Thurston. Called Len Thurston Ltd., the operation began in a two-car garage when Thurston, in the evenings, undertook the work his daytime employer turned away. The first major machinery purchased was a lathe, a power hack saw, and a welding machine: the price, $800. The seller was a farmer who had kept the equipment in an unused cattle stall. Today just one of the company's four tilt and load trucks would cost more than 100 times that amount; however, the original welding machine is still in use.

As the business expanded it moved to a shared facility on Rutherford Road that provided 1,500 square feet of floor space. Succeeding moves offered 4,000 square feet on Orenda Road and 14,000 square feet at the corner of Hansen and Rutherford roads. In 1970 John Thurston joined the firm; eight years later he purchased the company and changed its name to John Thurston Machine Limited. The latest move was to 98 Rutherford Road in 1980.

The new, 30,000-square-foot facility is equipped with four dock-level doors, one ground-level door, one overhead crane, and two overhead rails. The machine shop offers boring mills, lathes, and machines for milling, drilling, grinding, and broaching. The welding and fabricating shop includes a Pearson shear, press brake, plate rolls, cutters, grinders, welders, and other exotic equipment. These, and all the smaller equipment needed to supplement almost any job, are kept humming by 45 skilled employees.

Recent projects include converting a tube line from a linear to a circular conformation, post office carts, and a seed coater going to Finland. John Thurston Machine Limited's service is extensive and comprehensive.

John Thurston Machine Limited occupied this new, completely equipped metalworking facility in 1980. Located at 98 Rutherford Road, it is kept humming by 45 skilled employees.

CANADA PACKERS INC.

The opening of a highly specialized and ultramodern food service plant by Canada Packers Inc. in Brampton some 15 years ago positioned the company for accelerated participation in the growing "out-of-home" food segment.

Today the facility at 145 East Drive is an integral component of the corporation's nationwide food service operations. Production, sales, and distribution have exceeded original objectives, while plant and equipment have been expanded to state-of-the-art capability.

Processing of a few hundred items for the hotel, restaurant, and institutional trade commenced on November 13, 1972, in 50,000 square feet of manufacturing and office space on the 10-acre Brampton site. Costing $3.5 million, the plant incorporated the latest in structural design and production facilities. Start-up staff totalled 129 to handle approximately 2,000 food service accounts transferred from Toronto to Bramalea. By comparison, the work force is presently 300 (120 salaried and 180 hourly), and the customer base numbers 4,000.

The product line of fresh and processed foods has also broadened over the years to exceed 2,000 items, tapping the growth market created by demand for prepared seafood and entrées, snacks, and precooked finger foods by the burgeoning chain restaurant and fast-food industry. In addition, distribution has been increased to health care and industrial cafeterias and the airlines.

In response to steadily escalating volume, the Brampton operation has more than doubled its sales staff to 60 and currently deploys a 28-vehicle fleet—from 11 at the outset—to serve a market

The Food Service Division of Canada Packers Inc. in Brampton has expanded its facilities and product line to meet rising requirements for prepared foods in central Canada.

stretching from the Quebec border to Windsor, north to Moosonee, and west to Sault Ste. Marie. In the spring of 1985 Canada Packers invested $6.5 million in a 25,000-square-foot addition to manufacturing and storage areas, as well as new technology to produce precooked meats at the East Drive locale.

The corporation's food service operation at Brampton has justified many times over the timing and location of the venture. Officials of Canada Packers Inc. cite in particular the access to skilled labor and growing markets from the central vantage point of the city.

SKYLINE STEEL COMPANY

A Brampton steel specialist has recently accomplished a transition to public status in association with several related companies under a single organization capable of wide-ranging design, products, and service for the North American construction industry.

Skyline Steel Company was a well-established, private fabricator with a solid record in the industrial/commercial market when it joined forces in 1986 with Ennis Paiken of Hamilton, a producer of reinforcing steel (and subcontractor to the new Toronto domed stadium), and Ennis Steel of Port Robinson, a distribution firm. Coincident with the inclusion of Buffalo Structural Steel, the new public company known as Ennisteel Corporation was listed on the Toronto Stock Exchange of September 17, 1986. Within two months the alliance was further strengthened by the acquisition of Federal Pipe and Steel with distribution bases in Chicago and Detroit.

The formation of Ennisteel and the extension of coverage to the entire Great Lakes area constitute a major advancement for his firm in the view of Ted Boksa, who is the founder and president of Skyline Steel as well as executive vice-president of Ennisteel.

The Skyline shops, drafting rooms, and offices at 353 Clarence Street in Brampton will continue to respond to the fabrication requirements of a wide variety of structures, such as the recent Bramalea and Waterloo City Centre extensions, and the Noxell industrial plant in Mississauga for which more than 1,000 tons of steel was provided. In addition to a large volume of work in southern Ontario, Boksa, who is assisted by son Richard as project coordinator, has successfully bid on contracts as far west as Moose Jaw, Saskatchewan, and south to Fairfax, Virginia.

A native of Toronto, Boksa received a thorough grounding in all facets of steel design, fabrication, and erection during eight years of employment with Standard Iron and Steel prior to joining Dominion Structural Steel in 1958 and then United Steel three years later. He launched Skyline Steel Company on June 10, 1965, with a staff of six (it now numbers 42) at 535 Wilson Avenue in Toronto. In 1970 Boksa built a new plant and head office occupying 7,700 square feet on the present two-acre site where facilities have since been enlarged in several stages to cover 25,000 square feet.

Ted Boksa, founder and president.

Skyline Steel Company's plant and office at 353 Clarence Street.

BRAMPTON BRICK LIMITED

A strategy designed to increase productive capacity, improve operating efficiency, and expand the geographical scope of its markets will culminate in 1988, when Brampton Brick Limited occupies a modern, new manufacturing facility in the city.

The $28.5-million plant is under construction on a 100-acre site purchased for $4.6 million from Caterpillar of Canada Ltd. State of the art in design and technology, fully automated, and energy efficient, the project represents the final stage of long-term planning to consolidate the firm's position as the second-largest brick maker in Canada. Earlier moves contributing to the expansion program included the $3.9-million acquisition of the assets—and a strong presence in the Quebec market—from Brique Citadell Ltée in February 1986 and a decision by the family owners of Brampton Brick to make the company public later that same year.

Start-up of the new plant just north of the existing brick works early in 1988 will lead to full commercial production in May. An annual volume of 120 million architectural clay bricks—up from 59 million—will be attained at Brampton, while Citadelle output will add 42 million. The total effect of the transition will equip the company to meet rising demand from its large Eastern Canada market, which is predominantly in residential construction but also extends to institutional, commercial, and industrial building.

Brampton Brick products are available in approximately 70 colors and textures, including the premium antique-style John Price Brick, and have been supplied for such prominent structures as the Cyril Clark Library in Brampton, the Promenade Shopping Mall at Bathurst and Highway 7, and the Bay Charles Towers in downtown Toronto.

Historically, Brampton Brick Limited is one of the community's oldest continuing businesses, tracing its roots back to the establishment of Brampton Pressed Brick Limited in 1860. The business was purchased and given its present name in 1949 by Sam Rosenbaum, Manny Kerbel, Morris Kruger, and Allan Kerbel, whose families retain majority control and fill senior management roles. Directors and officers of Brampton Brick Limited are Barry Kornhaber, president; Samuel M. Kruger, vice-president/marketing; R. Michael Kerbel, vice-president/finance; Jeffrey G. Kerbel, secretary/treasurer; and Lloyd S.D. Fogler, Q.C., and Jean Fournier, who were appointed directors in conjunction with the public company designation.

The board of directors on induction day, November 14, 1986, Toronto Stock Exchange. Left to right are Lloyd S.D. Fogler, J.G. Kerbel, S.M. Kruger, B. Kornhaber, and R.M. Kerbel. Absent is J. Fourniere.

PARTNERS IN PROGRESS

BRAMALEA ANIMAL HOSPITAL

Professionals and experts in the provision of health care and treatment to companion pets function in a spacious, sophisticated setting at the ultramodern Bramalea Animal Hospital.

Bramalea Animal Hospital was first established in the early 1950s in a small building on Highway 7 in what was then rural Chinguacousy Township. Founder Dr. Robert Williams, who would later preside as the last township reeve prior to the 1974 advent of regional government, sold the practice in 1970 to Dr. Roger Footman. The first phase of the present hospital on one acre at 8640 Torbram Road was completed by Dr. Footman in June 1974. Subsequently, Dr. John R. Gordon, who had joined the staff the previous year, took over the practice, and early in 1982 he enlarged the structure from 2,100 to 5,300 square feet to cope with a rising case load that has since soared to five times the volume of the mid-1970s.

Successive expansions of staff, building, and equipment have occurred over slightly more than a decade under the guidance of owner Dr. Gordon, in step with the dramatic surge in city population. Today this full-service practice of veterinary medicine counts an annual clientele in excess of 15,000 canine, feline, and avian patients.

Dr. Gordon ascribes the success of his enterprise to a basic love for animals and a rapport with their owners, qualities that are shared by Dr. Othmar "Oats" Telep, a full partner in the hospital since 1983, and their associates. The facility employs three other veterinarians as well as two animal health technicians, three kennel attendants, one groomer, and five rotating receptionists to handle the myriad assignments arising from a broad range of health service delivery. By contrast, when Dr. Gordon bought the practice in January 1975, personnel consisted of an assistant veterinarian, a receptionist, and a kennel attendant.

Bramalea Animal Hospital, a full-service practice of veterinary medicine at 8640 Torbram Road, has grown since its inception in the early 1950s to accommodate a case load encompassing 15,000 canine, feline, and avian patients.

The current level of activity requires the deployment of two operating theatres, a laboratory, radiology room, three examination areas, two treatment areas, four large kennel wards, and an isolation room. In addition, Bramalea Animal Hospital maintains large boarding kennels inside, 15 outdoor runs, and a grooming parlor. A complete burglar alarm system, augmented by fire, heat, and smoke detectors, affords total protection to the premises. And in 1985 all functions from medical data to accounting were transferred to a computer system.

FLOWERLEA DAIRY LTD.

One of the largest independent dairies in the region is in glowing health at the completion of a quarter-century of forward progress accompanied by sweeping improvements to product, plant, and marketing.

Flowerlea Dairy Ltd. has developed a specialized clientele for its fluid milk production in an area reaching from Peel to Oakville and Toronto. The road to success had been rocky on occasion in the face of fierce competition from the chain and convenience stores, but the locally owned enterprise has prevailed by adapting to changes in technology, packaging, and distribution.

Flowerlea Dairy today is much more sophisticated in operation and capability than the small Brampton dairy purchased by a group of 20 district milk producers on January 2, 1962. The new owners brought to their acquisition a basic knowledge of dairying as well as experience in shipping milk as proprietors in Dairyland Transport Co-operative. For some time they had explored the prospect of an exclusively fluid milk dairy rather than accept lower prices for the quantities of their milk diverted to manufactured products.

Events moved swiftly after the inaugural meeting of the proposed dairy co-op at Snelgrove Hall on November 13, 1961, when the availability of the Pocock Dairy (established in 1920) was disclosed. A new name for the business, Flowerlea Dairy Co-operative, was adopted on December 4, 1961, reflecting Brampton's image as the national floral capital and incorporating the English word for meadow. By year-end legalities were completed for the takeover by the charter executive of Flowerlea, comprised of president Neil Armstrong, vice-president James KcKane, secretary/treasurer John Clarkson, and directors Everard Gray, Herbert Watson, and Russel Graham. The other founding members were: Alex Dean, Orval Hostrawser, Harold Wilson, Joseph Sherman, George Gardhouse, Newton Little, Edward Armstrong, Norman Armstrong, David Armstrong, Jack Wilkinson, Hunter Cation, Keith Parkinson, Clure Cation, Joseph Gray, and Sam Gray.

Innumerable extensions to the property, buildings, and equipment at the Flowerlea premises on Park Street in downtown Brampton have been effected over the past 25 years. Beginning in 1971, a conveyor system was installed to connect bottling, refrigeration, and loading docks, along with a concrete pad in front of the plant. Then the plant exterior was enhanced with aluminum siding, trim, eavestroughing, and new roofing. Properties north and south of the original site were purchased in the early 1980s, followed by the acquisition of 57 Nelson Street West in 1986.

A transaction of lasting value to Flowerlea was the purchase of the Bolton Dairy in February 1973, following which the assets were transferred to Brampton. Later that year the Bolton raw milk truck route was bought by Dairyland Transport. Both operations continue to service the Bolton area. As the volume of milk sales increased, Flowerlea discharged the mortgage held by the previous owner in 1977.

Internally, Flowerlea made the transition from co-operative to corporate status under its present name on February 7, 1980. And on March 23 it completed a five-year program of conversion to metric measurement for containers. This very busy period on the administrative side culminated in the purchase of a computer in December 1981; subsequently, a new Rexon computer was installed in 1985.

In the course of modernizing, the dairy has run the entire gamut of packaging—bottles, cartons, jugs, and plastic bags—until today it utilizes primarily 250- and 500-millilitre containers. Refrigerated equipment has been continually updated, including the fleet of trucks that had advanced from electric to gas to propane and diesel power. Plant capacity over 25 years has risen from 4,000 to 16,000 pounds per hour, while annual sales have increased from $650,000 to over $12 million.

The market for Flowerlea Dairy resides predominantly in the catering, restaurant, schools, variety stores, and the institutional sector with lesser, but substantial urban and rural home delivery volume. By contrast, retail outlets and door-to-door sales absorbed most of the production in the early years.

Throughout the years of change and growth a strong element of stability has existed in the stewardship of only two general managers. Al Dunsmore was the chief executive officer from inception until March 1967, when he was succeeded by Ezra Ginther.

Flowerlea Dairy Ltd. has supported a variety of community activities, including scholarships, sports, the annual Dairy Princess competition, and other charities.

Flowerlea Dairy Ltd. is much more sophisticated in operation and capability today than it was when opened in 1962. Compare the plant's exterior in 1962 (right) with 1987 (top), and the original bottling equipment (below left) wth today's packaging area (below right).

283

CFNY-FM

The once-shaky little community station is a creative and corporate success now, a high-flying phenomenon of Canadian radio acclaimed by cosmopolitan audiences, the music industry, and the City of Brampton.

CFNY-FM has staged a stunning recovery from the brink of financial ruin, barely seven years ago, to consummate broadcasting excellence of metropolitan magnitude and national stature. From studios at 83 Kennedy Road South, via a powerful transmitter in Toronto's CN Tower, a 35,000-watt signal sweeps from Georgian Bay over the country's most populous urban corridor and deep into New York State to captivate a weekly audience exceeding 500,000. This loyal, well-educated, and upscale following—a tenfold increase in numbers since 1980—contains the largest segment of $50,000-plus income earners of any radio competitor in the market.

The destiny of the obscure media outlet whose time had come was first perceived by David Marsden, director of programming and operations, who is credited by owner Selkirk Broadcasting Limited for the dramatic turnaround. In fact, Marsden almost presided over the demise of his employer after arriving at CFNY-FM's makeshift facility in an old house at 340 Main Street in 1977. Originally the local station was CHIC-FM, a late 1960s offshoot of CHIC-AM—"Where The Girls Are"—one of the first disco-style stations in the Toronto area. Both stations were founded and operated by Harry and Leslie Allan of Toronto; in 1974 CHIC-FM was renamed CFNY-FM, while CHIC-AM subsequently became the multilingual Brampton station CKMW.

When the shadow of receivership fell over CFNY-FM in 1980, interim management was assigned by the Canadian Radio and Television Commission to Civitas Corporation, a Quebec-based radio holding company. The Brampton operation survived and was acquired in 1983 by Selkirk Broadcasting, which proceeded to infuse stability, encouragement, and improvement. Within weeks of the transaction, Selkirk expended $145,000 on the tower transmitter, then followed with installation of the CFNY/Technique Showcase Studio on the observation deck of the skyline structure. As visiting rock bands and groups ascended the downtown locale for guest performances, the company stepped up sales efforts for CFNY-FM from offices at 60 St. Clair Street East in the fast lane of media promotion.

The "Spirit of Radio" theme permeating the Brampton assault on the air waves flows from Marsden's extensive experience in the front lines of such industry giants as CKEY-AM and CBC television. Landing in Brampton after a career break, he mixed rock in all its variations with a broad range of other music forms and the basic information elements, all the while fighting off despair as the station foundered. Marsden also

CFNY-FM had its beginnings in this house that contained the studios of its sister station CHIC. The house, at Two Ellen Street, is now owned by Flowerland. CFNY-FM later moved to 340 Main Street, and its studios and offices are now located at 83 Kennedy Road South.

assembled dedicated announcers and commentators, some of whom remain on staff after six to eight years and continue to contribute significantly to the winning format.

An outstanding achievement by Marsden and associates was the creation of the CASBY (Canadian Artists Chosen By You) Music Awards in 1981, a showcase for talent outside the mainstream of the entertainment world. Originally called the U-KNOW Awards, the venture attained both national recognition and popular acceptance during a coast-to-coast telecast by the CBC in 1985 from the Metro Toronto Convention Centre. The second annual CASBY presentation,

PARTNERS IN PROGRESS

featuring the return of co-hosts Carole Pope and Paul Shaffer but staged at The Kingswood Music Theatre in Canada's Wonderland, was underpinned by a national corporate sponsor in Heinz Ketchup. The awards were further enhanced by the participation of a full national radio and newspaper network, including stereo simulcast by the stations. In 1987 the CASBY gala will return to the Convention Centre.

The CASBY exercise is consistent with long-standing practices at CFNY-FM, which traditionally had outdone contemporaries in the introduction of new Canadian music and performers. Frequently, it has earned distinction as the first North American broadcaster to unveil rising young bands and vocalists. Coming full circle, the approach blends in a perfect fit between the music log and a discriminating audience.

Complementing the profusion of music, interspersing announcer preferences for rock, blues, jazz, and a steady stream of pre-releases, is a full-service station dedicated to informative substance as well as entertainment. CFNY-FM is anchored by a 10-person news staff that may editorialize on occasion, but generally delivers thorough reportage on current events, sports, weather, and traffic around the clock.

On the commercial side, advertising clients respond to a creative production team responsible for some of the most

The highest studio in the world, CFNY/Technics Showcase Studio is located on the observation deck of the CN Tower. It is the only radio station studio in the tower, which also is the site of the radio station's transmitter.

innovative copy in the business. Citations for writing, production, and sonic creativity adorn the walls at 83 Kennedy, alongside medals and awards won by the station in prestigious international competitions.

From a community standpoint CFNY-FM perpetuates a positive image of Brampton, in keeping with policy dating from the days of feeble 800-watt range to the emergence of the nation's hottest progressive station.

285

AMERICAN MOTORS (CANADA) INC.

The automobile assembly plant of the future is accelerating toward bold corporate objectives while topping the tank of economic growth in Canada's fastest-moving city.

American Motors (Canada) Inc. has crowned a presence of more than three decades in this country—and 27 years' residency in Brampton—with a $764-million investment in one of the most modern, highly efficient auto manufacturing facilities in North America. An aesthetically pleasing, single-level structure covering 2.2 million square feet, it represents the largest single commitment in the company's history.

The start-up of production in the superbly functional building in 1987 also signals the entry of AMC/Jeep/Renault (the signage reflective of the fifth-largest automaker in the world) into the intermediate car market. Appropriately, the new vehicle commanding all of the resources at the new plant is named the Premier—a creative accolade to automotive excellence and Brampton's favorite son.

Designed by Giorgio Giugiaro—recognized as the leading automobile designer in the world—the Premier will be crafted by some 3,000 highly trained workers in a plant that sets new standards of automobile technology in North America. Initial production is concentrated on a four-door sedan, to be followed by a two-door coupe in 1988. Production volume is scheduled at 75,000 units in the first year and 150,000 in year two of the assault on an upscale market identified with the Ford Thunderbird, Oldsmobile Cutlass, and Audi 5000.

The mammoth "E"-shape complex sits on 216 acres in the northeastern section of the Bramalea Business Park at Torbram Road and Williams Parkway. It incorporates the best features of automobile manufacturing plants all over the world to facilitate control over each function. The arms of the "E" accommodate body and paint shops, trim, and final chassis; high-tech services and processes throughout the plant ensure the utmost in preventive maintenance, quality resident engineering, and product support. Finally, the strength of AMC in Canada is underpinned by a dedicated work force drawn from experienced company employees and a selective hiring program.

An overriding positive factor contributing to the Premier is the Just-In-Time concept, an advanced approach to automaking with advantages in speedier turnaround, reduced inventory levels, and improved vendor performance. The system compresses ordering and receipt from outside sources to a few hours; incoming materials travel no more that 300 feet from point of entry via 37 commodity docks.

One of the integral parts of the Just-In-Time supply arrangement is a new

The new American Motors facility covers 2.2 million square feet, will employ 3,000 people, and begins production of the Premier on July 27, 1987.

$30-million satellite plant built by American Motors in Guelph. There about 500 employees are engaged in injection moulding of plastic instrument panels, assembled steering columns, and trim panels for shipment to Brampton. In all, up to 4,000 jobs could be created in the parts and components industries as a result of the Brampton project. And city authorities estimate that an overall 10,000 to 12,000 occupations in related manufacturing and supply will flow from the new AMC undertaking.

Breaking ground in the vehicle market is familiar territory for American Motors (Canada) Inc. From Rambler models at the inception of the first Brampton operation at Steeles and Kennedy Road in 1960, the company escalated to record output of 82,373 Hornets and Gremlins in 1974. A transition to Jeep manufacture in 1979 was later supplemented by the unique rear-wheel, four-wheel-drive Eagle passenger car. In 1986 the all-new Jeep YJ was added to Eagle lines at the original Brampton plant, which fills AMC's entire U.S. and Canadian quota of these vehicles.

The birth of the parent American Motors Corporation dates to 1902 in Kenosha, Wisconsin, where inventor Thomas B. Jeffrey assembled the first Rambler. During the first half-century the separate interests who would found AMC produced such classics as the Nash, Ajax, Lafayette, Hudson, Terraplane, and Essex. In 1954 AMC emerged from the union of Nash-Kelvinator and the Hudson Motor Company. Two years later the Canadian arms of Nash and Hudson were folded into American Motors (Canada) Limited,

The first Rambler completed by American Motors at the Kennedy and Steeles plant in Brampton, December 1960.

which soon embarked on re-location from inadequate Toronto premises to Brampton. The first Rambler built at Steeles and Kennedy rolled off the assembly line in December 1960.

Expansion in Canada has included the acquisition of the Holmes Foundry in Sarnia (engine blocks and castings) and of Canadian Fabricated Products in Stratford (soft trim). Internationally, American Motors bought the Jeep Corporation in 1970 and then negotiated a full partnership association with the giant French automaker Renault in 1979 to greatly broaden product line, technology, and sales exposure.

MAPLE LODGE FARMS

On the southwestern fringe of the city in a rural setting, one of the brightest stars in the Brampton industrial firmament is shooting for new horizons on the eve of nearly four decades of remarkable growth.

Maple Lodge Farms is a consummate Canadian business success story—a truly home-grown enterprise nurtured by the descendants of pioneer Ontario farmers into a world-class poultry-processing facility. And in 1987 the independent, family-owned company is consolidating stature as a dominant force in the Ontario chicken industry with the largest extension

The original barn on the May farm, circa 1908. The structure is part of the poultry operation today.

1930s Lawrence and Gwen May had established a dairy herd on the 180-acre property, which is located along the east side of Winston Churchill Boulevard just north of Steeles Avenue. Their sons Jack and Bob May acquired a basic knowledge of agriculture that would serve their aspirations in later years as the owners and managers of Maple Lodge Farms.

Lawrence May first established egg sales in the Toronto area during the 1930s.

The May family (left to right): Jean, Jack, Beth, Bob, David, and Larry.

Jack May first tested the marketing waters in 1947, when he bought a 1934 Chevrolet automobile, loaded it with eggs bought from neighbors, and sold door to door in Toronto. Two years later a trial run with 600 chicks led to a small venture in dressed poultry, followed by the adoption of the Maple Lodge name in 1950.

An early turning point in the fortunes of Maple Lodge Farms was the Swiss Chalet account secured in 1957. Current clientele includes restaurants of every size, such as Kentucky Fried Chicken and Scott

Jack and Bob May with staff in the late 1950s.

of productive capacity in its history.

The modern assembly of the latest techniques and technology responds to every facet of the proliferating chicken market, from consultation with chicken producers to computerized customer service. Research and development is combined with innovative flair in an ongoing program to introduce further processed chicken product lines.

The presence of the May family as proprietors and senior executives of Maple Lodge Farms on the present site dates to 1834, when their forebears cleared forests to grow crops and raise livestock on a frontier of limited opportunity. By the

Stores; specialty food service groups; hospitals; independent grocers; and all of the leading supermarket chains. In all, an extensive delivery fleet of specially equipped trucks supplies over 1,000 locations daily with both fresh and processed chicken products, while 29 live-haul trailers service more than 200 independent Ontario chicken producers.

A second pivotal move conferring major status on the Brampton firm occurred in 1965, when an addition of 15,000 square feet was accompanied by qualification for Federal Government inspection—a crucial prerequisite for chain store and consumer acceptance. Further extensions of 20,000 square feet in 1976 and 50,000 square feet in 1980 widened the product line, improved quality control, and increased output. Concurrently, overall sales and production soared dramatically for a 423-percent cumulative gain from 1974 to 1985 and a 78-percent increase over the past five years. In 1986 sales volume exceeded $135 million to rank Maple Lodge Farms as the 422nd-largest corporation in Canada.

The creation of a full range of 190 items of ice-packed and prepacked fresh poultry culminated in a resounding success in 1983 with the introduction of the

The first Norval United Church chicken barbeque, in August 1967, was started by Gwen May, Bob and Jack's mother.

company's popular Chicken Wiener. The product not only won widespread consumer acceptance, but also a Guiness Record for a 2,377-foot-long chicken hot dog produced at the Brampton plant. Upon the opening of its $10-million, 100,000-square-foot processing and distribution centre in 1987, Maple Lodge Farms will operate the largest chicken processing facility in the world.

Other factors contributing to leadership in the poultry industry—Maple Lodge Farms purchases more than one-quarter of all chicken grown in Ontario—arise from

Lawrence May carries eggs from the henhouse in the early days of operation.

a close working relationship with producers, inspection personnel, and customers to ensure continuity of quality, fresh product to the public. Truly a family-oriented business in every meaning of the word, the entire family is involved on a daily basis, ensuring first-hand awareness of the growing company's needs.

Responsibilities for sales and marketing are handled by Jack May, with his son, Larry, directing distribution and active in other areas of corporate interest. Jack's wife, Jean May, is director of payroll, while son-in-law David Grant heads up the sales team's daily activities. Jack and Jean have two daughters, Wendy Robson, who works in sales, and Susan Grant, who holds an active interest in the company.

Bob May is responsible for corporate administration, public relations, and production. His son, David, is director of plant operations, purchasing, and is also active in other areas of corporate interest. Bob's wife, Beth May, is director of personnel and is assisted by her daughters, Debi Kee, department manager, and Kathy Wienhold, in administration.

They supervise the various activities of more than 800 employees, all of whom contribute toward the continued success of the Maple Lodge family.

KANEFF PROPERTIES LIMITED

Kaneff Properties Limited builds properties for sale or rental. Its success can be measured in the value owners and buyers alike place on a property that has been "built by Kaneff." To date the firm has built 500,000 square feet of retail and commercial property, 6,000 detached homes, and condominium and rental apartment suites.

The history of Kaneff Properties Limited is more realistically the history of its founder, president, and owner, Ignat Kaneff. Kaneff arrived in Canada in 1951 from Bulgaria via Austria, speaking four languages—German, Russian, Bulgarian, and Croatian—but neither English nor French. He began working for G.S. Shipp and Son Ltd., first sweeping floors, then as a carpenter. Within seven months Kaneff had saved $300 for a down payment on a building lot. Borrowing the rest, he completed that first house, which Kaneff built mostly by himself. In 1953, using the house as equity, he began operations as an independent home builder, and by the following year was sufficiently confident to quit Shipp and hire two helpers.

In 1955 Kaneff built seven houses, and the following year incorporated his business under the name Kaneff Construction Ltd. The year 1957 saw the construction of the firm's first nine-unit apartment building, as well as an auto dealer's showroom and complex that got Kaneff into the automobile business in Brampton with what is now Frost Pontiac, Buick, Cadillac, Ltd., at 320 Queen Street East.

In 1961 Kaneff became a Canadian citizen and for the next 20 years concentrated his efforts, and flourished, in Mississauga. In 1982 he was named Businessman of the Year by the Mississauga Board of Trade. The company's first major projects in Brampton were the two highrise apartment buildings, located at 210 and 220 Steeles Avenue, followed by the 90,000-square-foot Bartley's Square mall in 1985. In 1987 projects under way in the city included the Stornwood Luxury Townhouses, 10 Steeles Plaza, and three highrise condominiums on Highway 10 called Pinnacle I, II, and III.

There is no question that the hard work and long hours have paid off for Ignat Kaneff. Today his firm has a staff of 200, operates out of a new building on Central Parkway West, and has assets that are conservatively estimated in excess of $300 million. At the same time he generously returns some of the benefits to the community that has allowed him to prosper. Kaneff has endowed an award for urban studies at the University of Toronto and runs an annual golf tournament to benefit the mentally retarded, collecting $150,000 to date, plus donations from Kaneff. On October 5, 1986, he assumed the chairmanship of the million-dollar Kaneff Charitable Foundation, an organization dedicated to the support of worthy causes, ranging from symphony orchestras to minor league hockey teams.

The Driftwood, a six-storey apartment building on June Avenue, was constructed by Kaneff Properties Limited in 1962.

The Pinnacle, a three-phase project, is a luxurious highrise condominium complex being constructed on Elgin Drive in Brampton.

STEARNS CANADA INC.

Stearns Canada Inc. is a manufacturer of nonwoven textile fabrics whose main ingredients are either rayon, polyester, or polypropylene. The firm converts fibres into web or sheet configurations or roll goods, of which 90 percent are supplied to others for conversion into finished products. President Ken Squires sees Stearns as the no-name company that turns out a product everyone encounters in one form or another, but nobody realizes it.

Stearns Canada traces its history back to The Stearns and Foster Company of Cincinnati, Ohio, founded in 1846. Its first Canadian plant was opened in Montreal in 1884. The Brampton plant got its start under the Union Carbide name and was acquired by Stearns in 1972, while the Trenton plant was built by the company in 1975 and expanded nine years later.

Having been a private operation since 1846, the company has undergone a few ownership changes in recent years. In December 1983 Stearns and Foster merged with the Ohio Mattress Company, and by September 1985 both the U.S. and Canadian textile divisions had been purchased by the Stearns Technical Textile Trust, which retains ownership today. Stearns Canada has an entirely Canadian management group, and the Canadian president reports directly to the Stearns Trust.

The Montreal plant produces primarily quilt batting and stuffing, while the Trenton operation concentrates on products used in disposable diapers, sanitary napkins, moist towelettes, household wipes, and fabric softener sheets. This product line is in keeping with the paper industry, where the nonwoven textiles business began. The Brampton plant contributes 35 percent of the firm's total Canadian sales, supplying a wide range of fabrics for furniture padding, automotive padding, and filtration systems. National General Filter Products, located next door to the Brampton facility, is such a good customer that there is an opening in the wall between the two plants, and a conveyer belt carries basic polyester medium for filters directly from Stearns with the simplest of guarantees on delivery.

Other major Stearns customers include Chrysler of Canada, which buys padding for doors; Procter & Gamble, a buyer of fabric that is then treated to become Bounce dryer sheets; and General Motors, where roll goods are delivered, cut up, and heat sealed to vinyl or cloth for use in automotive door panels. Stearns has recently begun to export product to markets in the United States, the Caribbean, and as far away as Australia.

The Stearns Canada Inc. staff, drawn mainly from Brampton, has remained stable at about 35 people since the plant was first opened.

Stearns Canada Inc. is a leading manufacturer of nonwoven textile fabrics. Shown is one of the Brampton facility's production lines.

Stearns Canada's Brampton facility contributes 35 percent of the firm's total Canadian sales.

POLY PLAST EDGEBANDING INC.

Poly Plast Edgebanding Inc. produces, inventories, and distributes a complete line of products for edgebanding decorative panels used in furniture and kitchen cabinets, as well as specialized sheet veneers, decorative mouldings, drawer sides, and specialized hot-melt adhesives. The edgebanding comes in polyester, real wood veneers, and PVC extrusions. Wrapall Mouldings Inc. is an associate company that bonds impregnated or saturated papers to assorted moulded substrates for use in the furniture trade. Poly Plast's assertion is that if you have a Canadian-made product using edgebanding, there is an 80-percent chance you are looking at its product.

Company president Ed Almeida was a 25-year-old Portuguese immigrant with a strong selling background when he founded Poly Plast in 1982. Operating with limited financial resources, he established a plant on Bram Court in Brampton, and by offering dramatically superior delivery schedules for polyester, wood veneer, and PVC edge banding, proceeded to capture a share of the furniture and kitchen trade in Toronto. At that time the firm consisted of Almeida and a single employee.

Early in 1983 Almeida formed a partnership with his former business associate Richard Locke, who became vice-president and secretary. The company incorporated that year, moved to larger premises on Bram Court, expanded sales to all of Ontario and Western Canada, and improved sales by a factor of five over the previous year. In 1984 a sales staff was added, producing sales expansion by a factor of eight and requiring additional office and plant staff.

The partners formed a new enterprise, Wrapall Mouldings Inc., which was incorporated in November 1985 to cater to the specialized moulding needs of the Canadian furniture industry. Poly Plast moved to a new plant at 12 Bram Court, engaged a sales agent in the United States to cover the eastern portion of the country, and increased sales by a factor of two.

By 1986 Wrapall required its own premises and 12,000 square feet were found on Bram Court. Additional machinery was imported from Europe, and a plant manager and six new employees were hired, bringing the total number of employees for both companies to 25. Poly Plast acquired sales agents and distribution centres in California, Florida, and the midwestern United States, and projects continued sales growth for 1987 and beyond.

Ed Almeida, president (seated), and Richard Locke, vice-president.

It is the policy of Poly Plast Edgebanding Inc. to promote from within, employ local people, and purchase from local Brampton suppliers, returning benefits to the community that has supported its rapid growth. Plans are under way for a new 50,000-square-foot combined facility.

RE/MAX BRAMPTON INC.—REALTOR

Staff of RE/MAX Brampton Inc.

RE/MAX Brampton Inc. is a firm of real estate brokers whose main business is the sale of residential properties. Of lesser but growing importance is the buying, selling, and leasing of factories and other commercial properties. RE/MAX is somewhat radical in the real estate industry because its sales people work on a 100-percent commission basis and, being a franchise operation, fewer expenses are incurred. Sales associates share fixed overhead expenses, pay for individual marketing expenses, and share in major marketing ventures in which they have a vote.

RE/MAX got its start in 1973 in Denver, Colorado, and moved into Canada five years later with an office in Alberta. RE/MAX Brampton was the company's first independent Ontario franchise, opening for business on April 1, 1980. Owner Ken McClenaghan began the business with one partner, increased the staff to 12 by year-end, and had 28 employees at the end of 1981. The following year a second office was opened at 44 Peel Centre Drive in Bramalea, and the total staff numbered 55. The two offices were split by the end of 1983, and RE/MAX became the single-largest real estate sales producer in Brampton.

The offices at 350 Rutherford Road South have been expanded five times—from an initial 1,100 square feet of floor space to 5,000 square feet today. The staff has increased to include 40 salespeople and the support personnel needed to keep them on the road.

By 1986 the RE/MAX franchise employed 14,000 people working from 1,010 offices. In Ontario there are currently 3,037 people operating from 155 offices. The RE/MAX people believe this is partly due to their policy of hiring only top sales producers, and then allowing them to earn more than as a conventional realtor. A RE/MAX sales associate can earn the equivalent of 77 percent of the sales commission, on average, as compared to only 58 percent with a conventional firm. The concept also seems to smooth out the cyclical highs and lows that have been a tradition in the real estate field.

RE/MAX Brampton Inc. is the only RE/MAX franchise that belongs to the Brampton and Toronto Real Estate Boards, and its manager, Jim Howell, is the 1987 Brampton Real Estate Board president. For the past two years he has led the firm's sponsorship of a combination summer auction/flea market on behalf of Cystic Fibrosis. Sales associates collect donations of goods from local business people, the landlord provides the use of the building and parking lot, and all auction and some market proceeds go to fight Cystic Fibrosis.

ZOCHEM
DIVISION OF HUDSON BAY
MINING AND SMELTING CO., LIMITED

Hudson Bay's new zinc oxide plant opened in 1974.

Zochem produces zinc oxide. That's all the company does, but it does it efficiently and well. Hudson Bay Mining and Smelting Co., Ltd., is the fourth-largest Canadian producer of zinc from its mine in Flin Flon, Ontario, and 20 to 25 percent of the mine's production comes to Brampton for conversion. Zochem melts the raw zinc, then heats it further until it becomes a vapor that is then exposed to air, oxidizes, and becomes zinc oxide. The concept is quite simple, but since zinc oxide is highly reactive as a vapor and considerable heat is generated by the oxidizing process, technological tricks are required to produce the various end qualities required.

Hudson Bay Mining began operations in 1927 in Flin Flon. Zochem was founded in 1933 by a group of French investors and financiers. Located in Montreal, the operation was originally called Zinc Oxide Company of Canada Ltd., and from day one has purchased its zinc requirements from Hudson Bay Mining.

From 1939 to 1945 vast quantities of zinc oxide were required for use in smoke bombs, stimulating considerable growth for Zochem. By the mid-1950s Zochem was sold to a Montreal group, and in 1965 was sold to Hudson Bay Mining.

That same year Hudson Bay Mining purchased 15 acres in Brampton and established Hudson Bay Diecasting, producing zinc alloys die cast specifically for the auto industry. The site at One Tilbury Court was central to the auto industry; it was located in a pleasant community kind to industry. The original plant used only six acres and expansion of it was possible by erecting mirror images of selected portions, and this was done later for a new plating plant. In 1973 a new plant was built for Zochem, which then moved in 1974 from the inefficient old plant in Montreal to Brampton as the equivalent of a division of Canadian Metals Division of Hudson Bay Mining and Smelting Co. Ltd. In 1984 Hudson Bay sold the die-casting operation, split the Brampton site roughly down the middle with 7.5 acres to each, and Hudson Bay Diecasting became a separate entity.

Zochem now has a plant capable of producing roughly 30,000 metric tons of zinc oxide annually. Production in 1986 was 22,000 tons, produced with a staff of 16. The single-largest use of zinc oxide is in the manufacture of tires. To make modern rubber, raw rubber is treated with sulfur, an agent used to accelerate the process, and zinc oxide, which activates the accelerator. There are approximately 250 million tires produced annually in North America, each using between seven and eight pounds of rubber, of which 5 percent is zinc oxide. Most people are more familiar with zinc oxide as one of the base materials in calamine lotion and baby creams.

THE BRAMPTON REAL ESTATE BOARD

The Brampton Real Estate Board of 1987 is greatly magnified in terms of roster, operation, and service from the small founding group of progressive-minded brokers and salesmen who associated in a quest for professionalism and co-operation more than 30 years ago. With 1,050 members, advanced office technology, and multiple listing property sales of $780 million in 1986, the board currently ranks about seventh among its Ontario contemporaries and is 15th in national standing.

By contrast, the inaugural meeting convened in the old Queen's Hotel on March 2, 1955, enrolled a total of 25 members in what was originally designated "The North Peel, Dufferin and Halton Real Estate Board." The area of coverage, including Bolton, Malton, Georgetown, most of Caledon, and surrounding rural districts, has remained intact from the outset, while the present name was adopted in 1957.

Of historical note, charter president Harold W. Knight continues to hold membership in the BREB. His fellow members on the first executive were Harold A. Clarke, honorary president; Harold S. Hare, vice-president; Samuel D.H. Reid, secretary/treasurer; and Arthur Cleverley and Howard May, directors.

For the first two years meetings were held in members' homes, with Gladys Knight as part-time secretary. In 1957 the board shared staff and office space with the Brampton Chamber of Commerce in the Dalent Building on George Street—at an annual cost of $265!

By 1970, as the pace of residential and industrial growth quickened, membership had increased to 230. Then it almost doubled over the next six years to 420 persons representing 70 real estate firms. Amid this surge of activity, the board built its attractive headquarters facility at 119 West Drive in 1975. There computerization of functions has been phased in, culminating with an on-line data base for instant access to listings by all members. A staff of six full-time employees currently handles the myriad details of member service, indexing, and cataloguing as well as special projects such as the weekly *Showcase*, a real estate tabloid published in co-operation with *The Brampton Guardian*.

The Brampton Real Estate Board supports selected civic projects of overall community benefit in the nature of a $24,000 pledge to the Peel Memorial Hospital expansion program in 1982 and recent donations to the Salvation Army and the Peel Children's Aid Society.

The Brampton Real Estate Board occupies these pleasant surroundings at 119 West Drive, Brampton.

MONARCH PLASTICS LIMITED

The rapid development of Monarch Plastics Limited as a leading manufacturer in the container field represents a highly successful entry into Canada by The Mehta Group, a worldwide industrial entrepreneurial organization with holdings on four continents.

In less than 15 years since acquiring control of the company, The Mehta Group has expanded production facilities, product line, market, and staff. A modern 63,000-square-foot plant located on four acres at 116 Walker Drive engages 90 to 100 personnel—up from 19 at the beginning—in the manufacture of plastic bottles and containers for a growing clientele of name-brand pharmaceutical, home care, personal care, and cosmetics companies.

Monarch Plastics was founded in Richmond Hill in 1972, but was operating in Downsview a year later when The Mehta Group bought control. The new owner moved the business to leased premises in Mississauga in 1975 and completed the purchase of 100-percent interest three years later in order to execute ambitious plans for advancement.

Coincident with construction of the existing plant in 1980, the firm installed an array of new equipment and technology while improving quality, service, and marketing. As manufacturing capacity was expanded by 75 percent to serve a much broader base of customers, the basic blow-moulding operations of Monarch Plastics were augmented in 1983 by heat-transfer equipment to decorate containers with lettering, designs, and graphics. Although its market is principally in southern Ontario, the company exports about 35 percent of its output to the United States. (The export volume is declining owing to the start-up of Monarch Plastics Inc. at Kenosha, Wisconsin, by the parent group.)

Upon completion in October 1980, the Monarch Plastics plant was one of the first south of Clark Boulevard in the Bramalea Industrial Park.

The Brampton site has proven ideal for the company, which was able to retain trained staff members who already lived in the area. In addition, adequate storage for a larger inventory and property for future expansion were secured.

Eric De Souza is president and chief executive of Monarch Plastics Limited, and also serves as regional director, U.S.A. and Canada, of The Mehta Group. Rui F. Pinto is vice-president/finance and administration, and Hemang D. Mehta, whose posting to Canada in 1985 reflected the group's objective of a strong presence in this country, is vice-president/business development. Harold Mastine has served as plant manager since 1979.

PATRONS

The following individuals, companies, and organizations have made a valuable commitment to the quality of this publication. Windsor Publications and The Brampton Board of Trade gratefully acknowledge their participation in *Brampton: An Illustrated History*.

Aircraft Appliances and Equipment Limited*
American Motors (Canada) Inc.*
Anthes Universal Limited*
Archdekin Funeral Home (1985) Limited*
Armbro Holdings Ltd.*
Armstrong Holdings (Brampton) Limited*
Attrell Toyota*
Aurora Engineering Industries Ltd. Osborn Tools Division
J.D. Barnes Limited*
Carole & Ron Beier
Bergman Graphics Limited*
Bramalea Animal Hospital*
Bramalea Limited*
Bramalea Personnel Incorporated*
Brampton Brick Limited*
Brampton Golf Club Ltd.
Brampton Hydro-Electric Commission*
Brampton Plumbing and Heating Supplies (1981) Ltd.*
Brampton Public Library and Art Gallery
The Brampton Real Estate Board*
Burlington Canada Inc.*
Canada Packers Inc.*
Canadian Tire Corporation Limited*
Canber Electric Inc.*
Carter Automotive Canada Ltd.
Caterpillar of Canada Ltd.*
Central Trust Company
CFNY-FM*
Chan's Auto Centre Ltd.
Chez Marie*

City of Brampton
Colony Lincoln Mercury Sales Limited*
Dachem Limited*
Dale & Morrow Insurance Limited*
Del/Charters Litho Inc.*
Dieco Manufacturing Ltd.*
Edward Graphic Supplies Ltd.*
Export Packers Company Limited*
 Max Rubenstein, Chairman
 Jeff Rubenstein, Vice Chairman
 John Hoburg Lee, President
FBM Distillery Co. Ltd.*
Flowerlea Dairy Ltd.*
Frost Pontiac Buick Cadillac Ltd.
Graff Diamond Products Limited*
Haessler-DeWay Ltd.*
Hare Real Estate (Peel) Ltd.*
Thomas Corkett Hendy
Huegenot Limited*
Humber Nurseries Limited*
Ideal Metals and Alloys of Canada Inc.*
InKan Limited*
Ivers-Lee Limited*
Kaneff Properties Limited*
Kenrod Welding & General Fabricating Limited
Knecht and Berchtold Ltd.*
Knox Martin Kretch Limited*
Kord Products Ltd.*
E.E. Lagerquist Central Sales Ltd.
Lansing Canada Inc.*
Lawrence Lawrence Stevenson*
LEPAGE'S LIMITED*
Life Rite Inc.*
Thomas J. Lipton Inc.
John Logan Chevrolet Oldsmobile Inc.*
Lorlea Steels Limited*
Lustro Steel Products*
McCleave Truck Sales Limited*
John McDermid, M.P. Brampton-Georgetown
Mactac Canada Ltd/Ltee*
M&P Tool Products Ltd.*
Manutec Steel Industries Ltd.*

Maple Lodge Farms*
The Marlatt Group*
E.A. Mitchell Limited Realtor*
Monarch Plastics Limited*
Nabisco Brands Canada Ltd. Grocery Division*
National General Filter Products Ltd.*
North American Decorative Products Inc.*
Northern Telecom Canada Limited*
Orion Sea Food & Steak House
Peel Mutual Insurance Company*
Peel Seed Growers Co-op
Poly Plast Edgebanding Inc.*
Pro Print Brampton*
Re/Max Brampton Inc.-Realtor*
Reynolds & Reynolds (Canada) Ltd.*
The Rice Group*
Richardson Greenshields of Canada Limited, Brampton Office
Rorer Canada Inc.*
Sherway Warehousing (1977) Inc.*
Skyline Steel Company*
John Victor Smith
Stearns Canada Inc.*
Stubbs & Massue Lithographers Limited*
Sunworthy Wallcoverings*
Texport, A Division of TNT Canada Inc.*
John Thurston Machines Limited*
Toronto Dominion Bank Dixie & Orenda Branch
Tresman Steel Industries Ltd.*
Voyageur Travel Insurance Limited*
Wellwood Containers Limited*
H.T. Wilson Insurance Services Ltd.*
Zochem
 Division of Hudson Bay Mining and Smelting Co., Limited*

*Partners in Progress of *Brampton: An Illustrated History*. The histories of these companies and organizations appear in Chapter 7, beginning on page 189.

BIBLIOGRAPHY

Lack of space restricts me from listing every source of reference used in the writing of this book, but the following, drawn from books, government reports, newspaper and magazine articles, and archival records, should provide additional information to those readers who require it. Most of these sources are readily available to the public.

1. Newspapers
Brampton Daily Times
The Guardian
The Conservator
The Globe and Mail
The Toronto Star

2. Special Collections And Government Records
Bramalea Consolidated Developments Ltd., Annual Reports—1964 and 1966.
Bramalea Consolidated Developments Ltd., *Bramalea, Canada's First Satellite City,* April 1958.
Bramalea Limited, Annual Report, 1985.
Bramalea Master Plan. Canadian Mitchell Associates Ltd., September 1969.
Brampton Alive. Annual directories published by the *Brampton Guardian,* Brampton.
Brampton Business Directory, The Brampton Board of Trade and the Business Development Office, City of Brampton.
Brampton Centennial. Brampton Chamber of Commerce, 1953.
Brampton Is Blooming. City of Brampton, Planning and Development Department, 1986.
Buffy's Corners, Today's Four Corners, A Look At Early Commercial Architecture. Brampton Heritage Board, 1985.
Building Images, An Architectural Review Of Brampton Past. Brampton Heritage Board, 1983.
Business And Finance In Ontario. Markham, Ontario: Business and Finance Publishers, Inc., 1985.
Bylaws of Brampton Mechanics' Institute. 1895.
Churchville, Reminders Of The Past. Brampton Heritage Board, 1984.
Dale Estates. Regulations of employment, circa 1948.
Dale Estates. Statistics.
Eclectic Female Institute, Brampton. Henry H. Hutton, Principal, 1863.
Haggert Brothers Annual Catalogue. 1874.
Huttonville Home & School Association. *The History of Huttonville School,* S.S. #1, Chinguacousy, from 1840.
Maps. Clerk's Department, Region of Peel.
Planning Information Report. Regional Municipality of Peel, Socio-Economic Analysis Section.
Records Of The County Of Peel, 1849-1973. Includes minutes, by-laws, reports, financial records, and records of associated boards and commissions.
Records Of The Village/Town Of Brampton, 1853-1973. Minutes, by-laws, reports, financial records, etc.
Reflections '85. Office of the Clerk, Region of Peel.
The Bramalea Story. Report by Close Brothers Ltd., Shell House, London E.C.2.
The History Of S.S No. 5 Chinguacousy, 1847-1962.
The Perkins Bull Collection. United Church Archives, Victoria University, Toronto.
Trade Talks. Brampton Board of Trade.
Yesterday Today. Brampton Heritage Board.

3. Books, Articles, and Theses
A History Of Peel County To Mark Its Centenary As A Separate County, 1867-1967. Brampton: The Corporation of the County of Peel, 1967.
Brampton's 100th Anniversary, 1873-1973. Brampton: The Corporation of the Town of Brampton and the Brampton Of Centennial Committee, August 1973.
Chapman, James T. "Study Of The Early History Of Education In Brampton: 1850-1880." B.A. thesis.
Clarkson, Betty. *At The Mouth Of The Credit.* Cheltenham: Boston Mills Press, 1977.
—————.*The Story Of Port Credit.* Toronto: University of Toronto Press.
Gagan, David. *Hopeful Travellers: Families, Land And Social Change In Mid-Victorian Peel County, Canada West.* Toronto: University of Toronto Press, 1981.
Guillet, Edwin C. *Early Life In Upper Canada.* Toronto: University of Toronto Press.
Harlin, E.A. and Jenks, G.A. *AVRO.* Shepperton, Surrey, England: Ian Allan Ltd., 1973.
Horn, Michael. *The Dirty Thirties.* Toronto: Copp Clark Publishing, 1972.
Hoy, Claire. *Bill Davis.* Methuen, Toronto: 1985.
Jeffreys, C.W. *The Picture Gallery Of Canadian History,* Vols. 1 & 2. Toronto: The Ryerson Press, 1945.
Jones, Peter. *History Of The Ojebway Indians.* Toronto: Canadiana House, 1861.
McDougall, R.L. *Our Living Tradition.* Toronto: University of Toronto Press, 1959.
McNaught, Kenneth. *The Pelican History Of Canada.* Markham: Penguin Books, 1976.
Miller, R.G. *The Bramalea Story.* Reprinted from the *Monetary Times,* no date.
"Ontario Revs Up With Auto Dollars." Maclean's, September 15, 1986
Peden, Murray. *Fall Of An Arrow.* Canada's Wings: Stittsville, 1979.
Pope, J.H. *Illustrated Historical Atlas Of The County Of Peel.* Toronto: Walker and Miles, 1877.
Proudlock, Michael. "Labour In Brampton." B.A. thesis.
Radcliff, the Reverend Thomas. *Authenic Letters From Upper Canada.* Toronto: The Macmillan Company of Canada Ltd., 1953.
Robertson, John Ross. *The Diary Of Mrs. Simcoe.* Toronto: The Ontario Publishing Company Ltd., 1934.
Smith, Philip. *It Seems Like Only Yesterday.* Toronto: McClelland and Stewart, 1986.
Tavender, Geo. S. *From This Year Hence.* A History of the Township of Toronto Gore, 1818-1967. Brampton: Charters Publishing Co. Ltd., 1967.
Trevelyan, G.M. *Illustrated English Social History.* Middlesex, England: Penguin Books, Harmondsworth.

Index

PARTNERS IN PROGRESS INDEX
Aircraft Appliances and Equipment Limited, 210-211
American Motors (Canada) Inc., 286-287
Anthes Universal Limited, 258-259
Archdekin Funeral Home (1985) Limited, 194
Armbro Holdings Ltd., 205
Armstrong Holdings (Brampton) Limited, 197
Attrell Toyota, 204
Barnes Limited, J.D., 240
Bergman Graphics Limited, 265
Bramalea Animal Hospital, 281
Bramalea Limited, 260-263
Bramalea Personnel Incorporated, 223
Brampton Board of Trade, The, 190
Brampton Brick Limited, 280
Brampton Hydro-Electric Commission, 268-270
Brampton Plumbing and Heating Supplies (1981) Ltd., 192
Brampton Real Estate Board, The, 295
Burlington Canada Inc., 199
Canada Packers Inc., 278
Canadian Tire Corporation Limited, 230
Canber Electric Inc., 239
Caterpillar of Canada Ltd., 252
CFNY-FM, 284-285
Chez Marie, 191
Colony Lincoln Mercury Sales Limited, 196
Dachem Limited, 236
Dale & Morrow Insurance Limited, 221
Del/Charters Litho Inc., 256-257
Dieco Manufacturing Ltd., 276
Edward Graphic Supplies Ltd., 232-233
Export Packers Company Limited, 272-273
FBM Distillery Co. Ltd., 202-203
Flowerlea Dairy Ltd., 282-283
Graff Diamond Products Limited, 274
Haessler-Deway Ltd., 250-251
Hare Real Estate (Peel) Ltd., 253
Huegenot Limited, 266
Humber Nurseries Limited, 241
Ideal Metals and Alloys of Canada Inc., 222
InKan Limited, 212-213
Ivers-Lee Limited, 208-209
Kaneff Properties Limited, 290
Knecht and Berchtold Ltd., 275
Knox Martin Kretch Limited, 220
Kord Products Ltd., 267
Lansing Canada Inc., 207

Lawrence Lawrence Stevenson, 271
LEPAGE'S LIMITED, 228-229
Life Rite Inc., 216-217
Logan Chevrolet Oldsmobile Inc., John, 254-255
Lorlea Steels Limited, 227
Lustro Steel Products, 193
McCleave Truck Sales Limited, 248
Mactac Canada Ltd/Ltee, 234
M&P Tool Products Ltd., 195
Manutec Steel Industries Ltd., 231
Maple Lodge Farms, 288-289
Marlatt Group, The, 238
Mitchell Limited Realtor, E.A., 198
Monarch Plastics Limited, 296
Nabisco Brands Canada Ltd. Grocery Division, 246
National General Filter Products Ltd., 200-201
North American Decorative Products Inc., 224
Northern Telecom Canada Limited, 242-243
Peel Mutual Insurance Company, 226
Poly Plast Edgebanding Inc., 292
Pro Print Brampton, 244
Re/Max Brampton Inc.-Realtor, 293
Reynolds & Reynolds (Canada) Ltd., 225
Rice Group, The, 264
Rorer Canada Inc., 218-219
Sherway Warehousing (1977) Inc., 235
Skyline Steel Company, 279
Stearns Canada Inc., 291
Stubbs & Massue Lithographers Limited, 214
Sunworthy Wallcoverings, 215
Texport, A Division of TNT Canada Inc., 249
Thurston Machine Limited, John, 277
Tresman Steel Industries Ltd., 237
Voyageur Travel Insurance Limited, 247
Wellwood Containers Limited, 245
Wilson Insurance Services Ltd., H.T., 206
Zochem Division of Hudson Bay Mining and Smelting Co., Limited, 294
GENERAL INDEX
Italic numbers indicate illustrations
A
Academia de Brampton, 38
Adams, J.O., 128
Alderlea, *53*
Algie, Harry, 108
Algie, William, 108

Allen, T.J., 184
American Motors (Canada) Ltd., *133, 142,* 144, *165, 169,* 170
American Revolution, 17, 20
Archdekin, Albert, 32, 33, *33*
Archdekin, Charlene, 32
Archdekin, Elgin, 32
Archdekin, Elmore, 32, *33,* 96
Archdekin, Evelyn, 33
Archdekin, Fennel, 32, 33, *33*
Archdekin, James, 32, 33, *33,* 163, 178, *179,* 182, 184
Archdekin, Ken, 33
Archdekin, Leo, 32, 33, *33*
Archdekin, Marlene, 32
Archdekin, Muriel, 32
Archdekin, Peter, 32, 33
Archdekin, Prudence, 32
Archdekin, Stan, 32, *33*
Archdekin, Thomas, 32
Archdekin, Viola, 32
Armington, Caroline, 92, *93*
Armstrong Brothers, 133
Armstrong's, Charlie, 97
Arts, 92, *93*
A.V. Roe Canada Limited, 133
B
Bailey, Bill, *61*
Ballantyne, Mr., 88-89
Barnhart, John, 51
Bay Horse Inn, 32
Beatty, Gordon, 101, 106, 184
Beck, J.S., 66-67
Bell, Alexander Graham, 64
Bell Telephone Company, 66, 68, 170
Black Creek Pioneer Village, *37*
Blain, Richard, 56, 109, 113, 115
Blain, Roswell, 100, 109, 115
Blain's Hardware, 56, 109-110
Bland, George, *28*
Bland, John, 36
Bloor, Ronald, 92
Boer War, 103
Book, Samuel, 90
Bovaird, David, 115
Bowles, Charles, 36
Boyle's Drug Store, *106*
Bramalea City Centre, *166-167,* 167
Bramalea Consolidated Limited, 161, 163, 164

Bramalea Limited, 161
Brampton: centennial, *173, 174, 175,* 176-178, *178,* 179, *179,* 182-183; county town, 24-25, 49-50; *flower town,* 109; inc., city, 176; inc., town, 51, 101, 176-177; inc., village, 41, 43, 49; naming, 40
Brampton and District Labor Council, 170
Brampton Board of Health, 90, 91
Brampton Board of Trade, 57, 58, 78, 88
Brampton Bowling Club, 94
Brampton Brass Band, 134
Brampton Central Public School, *85*
Brampton Cheerio Club, 117
Brampton Curling Club, 94
Brampton Driving Park, 72
Brampton Excelsiors, *95,* 98-99, 110, 180
Brampton Fair, 47-48, 72-73, *73, 74, 75,* 186
Brampton Firefighters Local, 170
Brampton Flying Club, 32, 96-97, *96-97*
Brampton High School, *82, 84,* 90
Brampton Hydro-Electric Commission, 32, 168
Brampton Intermediates, 91
Brampton Knitting Mills, 62, 64, *114,* 117, *151*
Brampton Mechanics Band, *76*
Brampton Model School, 90
Brampton police, 52
Brampton Pressed Brick Company, 104
Brampton School House, 77
Brampton Transit, 184
Brampton Welfare Society, 128
Brant, Joseph, 20, 21, *21*
Bristol, Richard, 21
British Arms Hotel, 90
Brown, R.K., 115
Brydon, William, 181
Buffy, William, 40
Buffy's Corners, 40
Bull, Bartholomew Harper, 124
Bull, Bartholomew Hill, 124, 126
Bull, Bartley, 124
Bull, Jeffrey, 124
Bull Dairy Farm, *123,* 124-125, 126
Bull, Duncan, 124
Bull, Harper, 134
Bull, Louis, 124
Bull, Perkins, 124, 126, *127*
Burlington Bay, 19, 21
Burnett, Henry, 62
Burrows, Fred, 94
C
Cabibonike, 21
Caesar, Lawson, *84*
Calder, Charles, *26*

Calvert-Dale, 109
Campbell, A.F., 89
Campbell, Allen, 56
Camp Kitchen Carnival Parade, *111*
Canada Company, 25, 29
Canadian National Exhibition, 72-73
Canadian Oil Company, 126
Capitol Theatre, 128
Capone, Al, 126
Carabram Festival, 186
Carnegie, Andrew, 87-88
Castlemore Golf and Country Club, 36
Castlemore school, 35
Centennial College, 173
Central Grammar and Public School, 82, 84
Central Peel Secondary School, *157,* 171
Centre Road, 36
Charters, C.V., *128,* 177
Charters, Harry, 115
Charters Publishing Company, 104, 129, 177
Charters, Reg, 115
Charters, Sam, 89, 100, 115
Chinn, Cecil, 177
Chinn, Florence, 177
Chisholm, Kenneth, 41, 50, *53,* 58
Christ Church, 64, 78
Cities and towns: Albion, 21; Bramalea, 36, 146, *160-161,* 161, 163, *163,* 164, *165, 166,* 167-168, 170, 183, 184; Caledon, 21, 49, 97, 184; Castlemore, 29, 31, 75; Cheltenham, 49; Chinguacousy, 17, 21, 23, 26, 29, 36, 37, *37,* 43, 44, 45, 46, 49, 50, 75, 85, 89, 98, 124, 161, 164, 167, 170, 183, 184, 186; Churchville, 39, 49; Coleraine, 29, 31; Cooksville, 49; Ebenezer, 31; Erin, 21; Etobicoke, 29; Grahamsville, 35; Huttonville, 39, 80, 170; Malton, 49, 50, 55, 133, 143, 161; Mississauga, see *Toronto Township;* Silver Creek, 49; Springbrook, 39; Stanley Mills, 35; Streetsville, 49, 50-51; Toronto, 11, 21, 23, 43; Toronto Gore, 21, 23, 26, 29, 31, 35, 36, 39, 43, 46, 49, 90, 124, 184; Toronto Township, 23, 25, 29, 35, 39, 170, 184; Tullamore, 29, 35, 49, *54,* 90; Vaughn, 29; -Woodhill, 35
Civic Centre, *146-147,* 167-168
Claireville Conservation Area, 36, *149*
Clark, Cyril, 183
Claus, William, 20, 21
Close Brothers Limited, 163
Cole, J.W., *50, 51*
Colleges of Applied Arts & Technology, 172

Common School Act of 1816, 31, 35
Common School Law of 1816, 80
Concert Hall, 116
Cooper, Jimmy, *69*
Copeland-Chatterson, 104, *105*
Copeland, R.J., 88
County Agricultural Society, 29
County of Peel Agricultural Society, 47-48, 72, 124
Crawford, James, 141
Creditview Road, 39
Crimean War, 48, 91
Cullen, Brian, *94*
D
Dale, Edward, 62, 108
Dale Estates, 62, *62,* 68, 108, *108-109,* 110, 128, 143-144, 170, 186
Dale family, *102-103*
Dale, Harry, 62, *107,* 108, 143-144, 187
Dale, Robert, *107*
Davis, A.G., 128
Davis, Grenville, 100, 171, 180
Davis, Kathleen, 178, 182
Davis, William, 171-173, *173,* 178, 179, 180-181, *180, 182,* 183
de la Brocque, Boucher, 13
de Lery, Chaussegras, 16
de Oliveira, Evangelista, 38
Dick, Alexander, 89
Diefenbaker, John, 181
Disease, 29, 36-37, 89-90
Dixie Cup, 144
Doctor's Home, *37*
Dominion Building, *63, 144*
Dougherty, Patrick, 35
Downs, W. Emerson, 68
Ducreaux (explorer), 13
Dufferin-Peel Separate School Board, 175, 176
Duggan, T.W., 100, *107,* 108
Duke of Edinburgh, 176, *178,* 179
Dundas, Henry, 19
Dundas Street, 19-20, 26, 29
E
Earl of Carlisle, 178
Early land speculation, 25, 40
Eccles, Marlene, *75*
Eclectic Female Institute, 85-86
Economy, 43-44, 46-48, 53, 55, 57, 59-60, 62, 68, 103-104, *105,* 106, 108 *108-109,* 109, 110, *110,* 111-112, 117, 120, *123,* 124-125, 128, 133, *133,* 143-144, 161, 164, 165, 167-168, 170, 184, 186-187
Education, 31, 35, 39, 55, 56, 78, 80-82, *82, 83,* 84, *84,* 85, *85,* 86, 171-173, 175-176

Eldorado Park, *69*
Electricity, 39, 43, 62-64, 168
Elliott Blacksmith Shop and Garage, *68-69*
Elliott, Frank, *45*
Elliott, Harry, *45*
Elliott, John, 40, 41, 46, 72, 75, 78, 187
Elliott, Mary, *45*
Elliott, M.M., 101
Elliott, Richard, 40, *45, 68-69*
Elliott, Walter, *45*
Elliott, William, *39,* 40, 117
Eseard, Murray, *94*
Excelsior Hose Company #1, 101
Excelsior Intermediates, 100
F
Fan's Garage, *123*
Farmer's Institute, 116
Farr, Bill, 96, 97
Faulkner, Hugh, 182
Faulkner, Isaiah, 49
Fenian Raids, 113
Fenton, Wm. James, 90
Finlay, G.W., 164
Fires and fire protection, *98-99,* 100, *100,* 101, 110-111, 167, *175*
First Baptist Church, 78
Floods, 11, 134, *134,* 135, *136, 137, 138, 139,* 140-141, 143; diversion channel, 57, *130-131,* 143
Flower Festival, *100, 176, 177,* 186
Ford Motor Company, 167
Forster, J.W.L., 92
Fort Frontenac, 13
Fort Michilimackinac, 17
Fort Toronto, 13
Foster, W.J., 128
Four Corners, 39, 40, *144,* 186
Franco-Prussian War, 91
Free Presbyterian Church, 55
Frost, Leslie, 141
Furness, Ed., 128
G
Gage Park, 78, 134, *155,* 177, 178, 179
Gage, William, 78
Galbraith, W.J., *84*
German Canadian Club, 38
Gimby, Bobby, 178
Golding, James, *47,* 101
Golding's bakery, *35, 47*
Gore Road, 26, 29
Gorman, Terry, 170
Go-Transit, *173,* 176, 182, *182,* 183, *183*
Grace United Church, 78, *79, 177*
Graham, George, 35
Graham, Thomas, 35
Graham's Inn, 35
Grant, Simon, 35

Graves, Sidney, 115
Graydon, Gordon, 134, 181
Great Depression, 111, 125, 128
Great War Veteran's Association, 117
Guardian Angel's Church, 78
Guillet, Edwin, 71
Gullen, Barry, *94*
Gummed Papers Ltd., 104
H
Haggert Bros., *48,* 49, 56, 58, 59-60, 62, 86, 94, 104, 108
Haggert, John, 56, *58,* 59-60, 62, 63, 77, 86, 94, 187
Haggertlea, 60, *60-61,* 77, 94
Halton Rifles, 113
Hamilton, R., 50
Hansa Haus, 38
Harmsworth, David, 110
Harmsworth, Earl, 115
Harmsworth, Elmer, 110
Harmsworth, James, 101, 110, *110,* 115, 117
Harmsworth, Jim, 110
Harmsworth Paint and Wallpaper, 109, 110, *110*
Harmsworth, Russell, 115
Hartley, R.A., 50
Hawthorne Lodge, 124
Health care, 89-91
Heart Lake, 101
Heggie, David, 47, 90-91
Heggie, D.C., 130
Henderson, John, 50
Henderson, Murray, *94*
Heritage Complex, 186
Herkes, Andrew, 52
Hewetson, A. Russell, 110, 128
Hewetson, J.W., 128
Hewetson, John, 110, 128
Hewetson Shoe Company, 110-111, 128, 143, *143,* 171, 186
Hewetson, Vera, 170
High School Wasps, 94
Hiscox, R.J., 129
Holmes, John, 41
Home District Agricultural Society, 40
Houek, Jack, *73*
Houston's, *151*
Howland, Peleg, 41
Howland, William, 41
Hoy, Claire, 172
Hunter, Ida, 115
Hurontario Plank Road Company, 44
Hurontario Road, 20, 26, 36, 44, 51
Hurricane Hazel, 143
Hurst, John, 52
Hurst, William, 55

Hutton, J.P., 39, 50, 64
Huttonville Dam, 39, 64, 130
Huttonville Park, 130
Huttonville School, 80
I
Immigration, 26, 32, 38, 144
Imperial Optical, 144
Indian Council Ring, 21
Indians: Algonquian, 12; Chippewa, 11-12; Iroquois, 11; Mississauga, 11, *11,* 12, *12, 14,* 16-17, 18, 19, 20, 36-37, 98; Mohawks, 20; Ojibwa, 12, *18;* treaties, 19, 20-21, 36
Indian Village, 12
Industry: agriculture, *14, 15,* 36, 37, 39, 40, 43, 47-49, 59, 62, 72-73, 117, *123,* 124, 133-134; farm equipment, 49, 59; flower, *62,* 108-109, *108-109,* 186; fur, 3, 16-17, *16;* mills, 29, *30,* 31, 39, 40, 43, 53; potash, *35,* 40
International Tobacco Workers Union, 170
International Typographers' Union, 170
J
Jameson, Mrs., 71-72
Johnston, E., *84*
Johnston, William, 89
Jones, Augustus, 19, 37
Jones, Peter, 19, 37
Junior Bramalea Lions Club, *164*
K
Kawahkitaquibe, 21
Keats, Frank, 52
Kee, Lily Mae, 75
Keene, L., 130
Kellam, Rankin, 97
Kelley, Harry, 175
Kennedy, Tom, 97, 181
Kenney, Samuel, 40
Kidd, George, *94*
King George III, 20, 21
King James I, 178
Kirkwood, J.C., *84*
Kirkwood, W.A., *84*
Kitchens of Sara Lee (Canada) Limited, 167
K-Mart, 167
Knowles, E.W., 67
L
Laidlaw, George, 51
Lake Ontario, 11, *11,* 17, 19, 20
Lawrence, Elisha, 29, 31
Lawson, Ray, 40
Lawson, William, 40, 75, 78, 187
Ledlow, B., *84*
Lee, George, 98
Lees, R., *84*
Leidy, Minnie, 78

Letty, Private, 115
Lewis, J., 44
Libraries, *167,* 168, *172;* Mechanics Institute, 86-87; Carnegie, 78, *86,* 87-88, 168
Lindsay, John, 90
Lindsay, Rebecca, 90
Lorne Scots, 113, *113,* 115, 130, 177, 178
Lorne Scots Pipe Band, 134
Lougheed, Mr., *84*
Loyalist settlement, 17, 19-20, 21, 23, *23, 24, 24,* 25, *25,* 26-27, 29, 31, 35-37, 39-41
Lynch, John, 23, 36, 41, 44, 51, 78, 84-85, 88, 187

M
McCleave, James, 96
McClelland, Alex, 115
McClintock, T.A., 115
McClure Funeral Home, 78
McClure, George, 116
McCollum, Robert, *27*
McCombe, Peter, 170
McConnell, Norm, 100
McConnell, William, 72
McCraken, James, 130
McCulloch, Robert, 116
McDermid, John, 178, 185
Macdonald, Molly, 178
Macdonald, Ross, 178
McDowell, W.G., 171
McKechnie, Archie, 141
Mackenzie, William, 39
McKinney, Emmerson, 177
McMurphy, John, 64
McVean, Alexander, 29, 35
McVean, Alexander, Jr., 29
McVean, Archibald, *28,* 29
McVean, Helen Gordon, *28*
McVean, Peter, *27*
Magee, C.R., 141
Magnet Hotel, 35
Magrath, Thomas, 25-26
Mahaffy, William, 47
Mark Twain Society, 126
Marquis of Lorne, 53, 113
Meadowland Park, *149, 159*
Melvin, David, *167*
Merchant's Bank, 91
Meredith, John, 92
Mills, 29, *30,* 31, 39, 40, 43, 53
Milner, W.E., 100
Minto, Lord and Lady, 68
Mississauga Tract, 19, 20, 21
Moodie, Susanna, 29
Moore Dry Kiln, 144
Moorehead, Tom, 130

Morton, Adam, 82
Motorola Information Systems, 167
Mullin, John, 91
Municipal Council of Brampton, *41*
Murray, Alexander, *84*

N
Naismith, James, 91
Newmann, Susan, *174*
Newspapers, 88-89, 128-129; Banner and Times, 129; Brampton Progress, 89; Brampton Times, 47, 51, 55, 62, 84, *87,* 88, 89, 170; Christian Guardian, 78, 81; Conservator, 44, 56, 63, 64, 80, 88, 89, 94, 112, 115, 117, 120, *125,* 129, 130, 134, 170; Daily Times, 88, 178, 184; Globe and Mail, 89; Guelph Review, 129; Mercury, 88; New Toronto Advertiser, 129; Peel Banner and General Advertiser, 88, 89; Peel Gazette, 129, *129;* Port Credit Weekly, 129; Toronto Globe, 89; Weekly Standard, 88; West Toronto Weekly, 129; Weston Times and Guide, 129; Woodstock Sentinel Review, 89
New Survey, 21, 23, 25, 36
Niagara Escarpment, 11
Nichols, M.B., *73*
Nitty Gritty Brama Ching Wing Ding, 186
Northern Telecom, 167, *168,* 169
Norton, Theophilis, 35

O
Old Countrymen's Club, 117
Old Jail, 186
Old Timers Hockey Team (1966), *94*
Olympic Portuguese Club, 38
OPP Training School, 97
Orangemen, 39
Orangeville and Brampton Stage Line, 44
Ostrander, Wm., 80

P
Page Bros., 144
Pagitaniquatoibi, 21
Parker, F.A., 91
Pattulo, Alexander, 47
Peaker, Ken, *84*
Pearson International Airport, *132,* 133, 164
Peel Agricultural Society, 56
Peel County, separation from York, 24, 49
Peel Heritage Complex, 96
Peel Memorial Hospital, 116, 117-118, *118*
Peel Multicultural Council, 38
Peel plain, 11
Peel Village, 124, 126
Perkins Bull Convalescent Hospital, 126
Pickard, Archibald, 40
Pilkey, M., *84*

Pine and Rose Festival, 186
Plunkett Report (regional government), 184
Population, 37, 39, 40, 46, 51, 53, 57, 103, 144, 167, 168, 179, 183, 186
Pratley, Daniel, 44
Price, Norman, 92
Proudlock, Michael, 104
Prouse, Russell, 163
Purdue, Michael, 49

Q
Queen Elizabeth, 33, 108, 176, *178,* 179, 182
Queen Victoria, 53
Queen's Hotel, 64, *104*
Queen's Rangers, 19

R
Raffeux (explorer), 13
Raike, Stanley, 52
Railroads, 43, 44-46, 51, 53, *54;* Canadian National, *46,* 51, *141;* Credit Valley, 51, 53; Go-Transit, *173,* 176, 182, *182, 183, 183;* Grand Trunk, 44, 45, 46, 51, 53, 89; Toronto and Guelph, 44, 45
Recreation and entertainment, *70-71,* 71-73, 75, 91, 94, *94, 95,* 96-97, *96-97,* 98-100, *148, 149, 155, 156, 158, 159,* 186
Reeve, W.B., 41
Regional government, 183-184
Region of Peel, creation, 25, 176, 183-184
Region of Peel Museum and Archives, 186
Religion, 53, 55, 75-78, *82;* Anglican, 35; Baptist, 55; Catholic, 35, 77; Church of England, 53; Methodist, 19, 26, 37, 55, 75, 76, 77-78; Methodist Episcopal, 55; Presbyterian, 76; Primitive Methodist, 40, 55, 75, 78; Protestant, 35, 77
Rice Construction Group, 185
Rivers: Credit, 11, 12, 13, 16, 18, 21, 39, 65; Etobicoke, 11, 16, 19, 20, 21, 23, 94, *119;* diversion channel, 57, *130-131;* Grand, 20; Humber, 11, 13, 16, 29, 36; St. Lawrence, 13
Robarts, John, 173, 181
Robertson, Bob, *94*
Roman Catholic Separate School Board, 175
Ronald, William, 92
Rose Villa, *31*
Roselea Park, 134, 177, 178
Round Chain Company, 144
Royal Agricultural Winter Fair, 73
Russell, Peter, 20
Ryerson, Egerton, 81-82, 86, 171

S
St. Andrew's Presbyterian Church, 78, *80*
St. Anne's School, 38, 175

St. Francis Xavier School, 175
St. Joseph's School, 175
St. Mary's Roman Catholic Church, 78
St. Mary's school, 38, 175
St. Paul's United Church, 78, *78*
St. Thomas Agricultural Works, 60
Salisbury, Martin, 39, 40
Salisbury's tavern, 72
Salvation Army, 78, *81*
Salvation Army Band, 134
School Act of 1850, 81-82
Scott, John, 36, 40, 82
Seller, Thomas, 89
Senator Drum and Bugle Corps, 178
Seven Years' War, 17
Sharp, W.R., 67
Sheet Metal Workers union, 170
Shenich's shoe shop, *35*
Sheridan College of Applied Arts and Technology, *157,* 173, 175, 184
Sherwood, Reuben, 29
Shoppers World, 178
Simcoe, Elizabeth, 18, 20
Simcoe, John, 18-19, *19,* 23, 24
Sleightholm, James, 35-36
Smith, Robert, 49, 50
Smith, Sid, *94*
Smithers, Dave, 106
Snell's Lake, 101, 118
Stark, J.W., 117
Stevens, Stiles, 52
Stork, Christopher, 50
Stork family, *69*
Stork's drugstore, *35*
Strickland, Samuel, 29
Sullivan, C.W., 175
T
Tavender, George, 75, 90
Telephone service, 43, 62, 64, *65,* 66-67, *66-67,* 68, *121, 122,* 124, 170
Temperance, 55-56
36th Peel Battalion, 113
36th Peel Regiment, 113
Thomas J. Lipton Limited, 167
Timbrell, Dennis, *167*
Todd, Wesley, 46
Torbulglen, 29
Total Abstinence Society, 55
Transportation: auto, 43, 68, *104,* 120, 124, 163; bridges, 37; bus, 184; railroads, see *Railroads;* rivers, 11; roads, 19-20, 25, 26, 37, 39, *42,* 43, 76, *119,* 167
20th Halton Battalion of Infantry, 113
20th Lorne Rifles, *112*
Tye, George, 89
U
Union Metal Company, 144

Unions, 104, 170-171
United Auto Workers, 170
United Glass and Ceramic Workers union, 170
United Presbyterian Church, 55
Upper Canada, Lower Canada, formation, 17-18
V
Victory Aircraft Limited, 133
Vivian, Gord, 130
W
Wagg, Lewis, 110
Wagg, Lloyd, 110
War of 1812, 17, 21
Water supply, 101, 118, 168, 170
Watson, Garnet, *77*
Watson, John, 55
Weggishgomin, 21
Whillans, Ken, 6, 186, 187
White Spruce Park, *149*
Williams, Algernon, 66, 67
Williams, Terry, 163
Williams Shoe Company, 104
Wilson, Isaac, Mrs., 90
Women's Institute, 116, 117
Wood, P.L., 64
Woodill, Jane, 90
Woodlands Golf and Country Club, 36
World War I, 103, 104, 109, 111-113, 115, 117-118, 120
World War II, 32, 38, 103, 111-112, 129-130, 132-134
Wright, Dame (teacher), 80-81, 171
Wright, George, 49, 88-89, 171, 187
Wylie, Byrnell, *74*
Y
"The Young Bachelors," 75

THE BRAMPTON BOARD OF TRADE PAST PRESIDENTS

Dennis F. Cole	1986-87	Cecil Chinn	1961-62
Robert Bell	1985-86	Harold Knight	1960-61
Bruce Carruthers	1984-85	Lloyd Denby	1959
Keith Coulter	1983-84	Stan Stonehouse*	1958
Earnie Mitchell	1982-83	William Coupar*	1957
Martin Hughes	1981-82	C.G. Patterson	1956
James A. Phair	1980-81	Frank Richardson	1955
Terry R. Champ	1979-80	J.R. (Joe) Racine	1954
John H. Logan	1978-79	William M. Watson	1953
Donald R. Crawford	1977-78	Emmerson McKinney	1952
Harry R. Lockwood	1976-77	Wm. M. Robinson	1951
A.D.K. MacKenzie	1975-76	Cecil Carscadden	1950
G.W. (Joe) Harley	1974-75	F. Gordon Umphrey	1948-49
Watson Kennedy	1973-74	James Martin	1922-
Peter Montgomery	1972-73	G.W. McFarland*	1919-20
Hank Sawatsky	1971-72	F.W. Wegenast*	1917-19
R.C. Harvey	1970-71	R.H. Pringle*	1914-17
Richard Boyle	1969-70	J.H. Boulter*	1911-14
Ronald Rider	1968-69	G.L. Williams*	1907-11
J.A. Carroll*	1967-68	E.S. Anderson*	1907
S.G. Eisel	1966-67	T.W. Duggan*	1906-07
Gordon Vivian	1965-66	J.H. Boulter*	1904-06
Douglas Brown*	1964-65	E.O. Runians*	1890-96
S. Charters	1963-64	K. Chisholm*	1887
Edward Ching*	1962-63	*Deceased	